D1566629

The Last of the Great Observatories

THE LAST OF THE GREAT OBSERVATORIES

Spitzer and the Era of Faster, Better, Cheaper at NASA

George H. Rieke

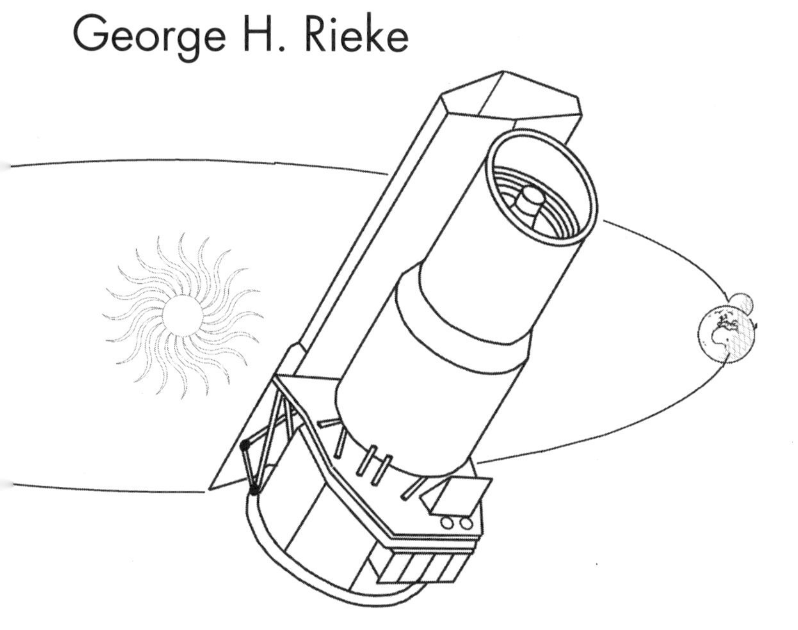

THE UNIVERSITY OF ARIZONA PRESS
Tucson

The University of Arizona Press

∞ This book is printed on acid-free, archival-quality paper.
Manufactured in the United States of America
11 10 09 08 07 06 6 5 4 3 2 1

Library of Congress Cataloging-in-Publication Data
Rieke, G. H. (George Henry)
The last of the great observatories : Spitzer and the era of faster, better, cheaper at NASA / George H. Rieke.
p. cm.
Includes bibliographical references and index.
ISBN-13: 978-0-8165-2522-5 (acid-free paper)
ISBN-10: 0-8165-2522-6 (acid-free paper)
ISBN-13: 978-0-8165-2558-4 (pbk. : alk. paper)
ISBN-10: 0-8165-2558-7 (pbk. : alk. paper)
1. Space Infrared Telescope Facility (U.S.) 2. United States. National Aeronautics and Space Administration—Management. 3. Astronomical observatories—United States—History—20th century. 4. Infrared astronomy—United States—History—20th century. I. Title.
QB82.U62S63 2006
522'.683—dc22

2005028811

Publication of this book is made possible in part by NASA contract 960785, issued by JPL/Cal Tech.

Dedicated to
Giovanni Fazio, Jim Houck, Mike Jura, Frank Low,
Mike Werner, Ned Wright, and Erick Young

CONTENTS

ILLUSTRATIONS

ACRONYMS

AAS	American Astronomical Society
AGN	active galactic nuclei
AXAF	Advanced X-ray Astrophysics Facility
CDR	Critical Design Review
CEO	chief executive officer
COBE	Cosmic Background Explorer
CRAF	Comet Rendezvous–Asteroid Flyby
DR	discrepancy report
DSN	Deep Space Network
EIRR	External Independent Readiness Review
EOS	Earth Observing Systems
EPI	Error Prevention Institute
ESA	European Space Agency
FBC	faster, better, cheaper
FY	fiscal year
GPS	Global Positioning System
GTO	guaranteed-time observer
HQ	headquarters
HST	Hubble Space Telescope
IAR	Independent Annual Review
IOC	in-orbit checkout
IPAC	Infrared Processing and Analysis Center
IRAC	Infrared Array Camera
IRAS	Infrared Astronomy Satellite
IRS	Infrared Spectrograph
ISO	Infrared Space Observatory
IST	instrument support team
JPL	Jet Propulsion Laboratory
JWST	James Webb Space Telescope
KSC	Kennedy Space Center

LMA Lockheed Martin Astronautics in Denver
MCO Mars Climate Orbiter
MER Mars Exploration Rover
MIC multiple instrument chamber
MIPS Multiband Imaging Photometer for Spitzer
MPL Mars Polar Lander
NAR Non-advocate Review
NASA National Aeronautics and Space Administration
OAST Office of Aeronautics, Science, and Technology
OMB Office of Management and Budget
OMV Orbital Maneuvering Vehicle
OSS Office of Space Science
OSSA Office of Space Science and Applications
PDR Preliminary Design Review
PI principal investigator
PMC Program Management Council
PNAR Preliminary Non-advocate Review
RCS reaction control system
SDIO Strategic Defense Initiative Organization
SIRTF Space Infrared Telescope Facility
SOFIA Stratospheric Observatory for Infrared Astronomy
SSC Spitzer (or SIRTF) Science Center
STL System Test Lab
TAR test anomaly report
WFPC Wide Field Planetary Camera 2
WIRE Wide Field Infrared Explorer
WPA Works Progress Administration

PREFACE

Building a major space science mission is one of the most challenging of human activities, beginning with the need to persuade other humans to commit substantial resources to the effort, then overcoming myriad technical and managerial problems, and finally surviving the risky ordeal of launch and operation in space. It is high adventure in a realm of obstinate politicians and recalcitrant technology. When everything works out, the results are exhilarating; otherwise no one would do it. This book tells the story of one such adventure from my perspective as a participant. So that I can concentrate on the story without digressions, I discuss technical details and assess management approaches in appendices.

I tell intertwined stories connected with the mission to launch the Space Infrared Telescope Facility (SIRTF), now known as the Spitzer Telescope. Spitzer is the last of the four "Great Observatories" to be completed. The Great Observatories program, which comprised Spitzer, the Hubble Space Telescope, the Chandra X-ray Observatory, and the Compton Gamma Ray Observatory, was designed to advance our understanding of astronomy from the infrared to gamma ray spectral ranges.

I begin with the process of building the observatory, combining that story with the story of NASA Space Science through two turbulent decades. I have interviewed many participants in the overall space science effort, and their words will help to build a picture of the sometimes chaotic political and managerial processes by which missions are conceived, approved, and developed. Other material was obtained in a variety of public meetings such as reviews, group meetings associated with the project, and NASA advisory committee meetings. These events took place over the entire duration of the SIRTF mission, but from the beginning I made a habit of writing down particularly telling quotations. In all cases I have reproduced the discussions to the best of my ability based on notes taken at the time. This is also the story of a project built at the peak of NASA's "faster, better, cheaper" experiment with streamlining and downsizing its mission development process. By the time SIRTF had been completed, a number of mission failures had undermined faith in faster, better, cheaper, and a more conservative approach was imposed. The story therefore examines the strengths and weaknesses of this discredited management style and their ultimate impact on future missions.

I have written the story from the perspective of a participant rather than as a disinterested and distant observer. Thus, I record the adventures and reactions of an initially (and perhaps still) naïve university professor encountering and occasionally influencing a series of events that led to the launch of a powerful new astronomy mission.

Many people contributed to the process of writing this book. I thank Lee Armus, John Bahcall, Steve Battel, Chas Beichman, Nancy Boggess, Larry Caroff, Vasilis Charmandaris, Lennard Fisk, Art Fuchs, Dave Gallagher, Dan Goldin, Bill Green, Roger Grubic, Wes Huntress, Steve Isakowitz, Kevin Kelly, Tim Kelly, Chris King, Johnny Kwok, Frank Low, Frank Martin, Gary Pace, Charlie Pellerin, Marcia Rieke, Larry Simmons, Domenick Tenerelli, Nick Vadlamudi, Phil Wagner, Dan Weedman, Debbie Wilson, Bob Wilson, and Erick Young, among others, for insightful interviews. I am also indebted to Whitney Clavin, Dave Gallagher, Gerry Neugebauer, Marcia Rieke, Larry Simmons, Robert Smith, Mike Werner, and Brooke White for reading the manuscript and suggesting many improvements. I thank Frank Low for reproducing his first sketch of the SIRTF warm launch concept. I also thank Dave Bearden for running his cost/complexity model on SIRTF. Finally, I thank Melinda Conner for a careful and thorough editing of the original manuscript.

The Last of the Great Observatories

1

FRIDAY THE THIRTEENTH

■ NASA released the Announcement of Opportunity to build an instrument for the Shuttle Infrared Telescope Facility (SIRTF) on Friday, May 13, 1983. No one heeded the omen. I assembled a superb team who spent a number of months writing contributions to a proposal. With help from my astronomer wife, Marcia, I assembled the contributions on a primitive "luggable" word-processing computer. We worked nights, days, and weekends. Although it was a pleasant walk from home to our offices at the University of Arizona, we drove back and forth so the computer could stay with me and I could keep massaging the words it carried. Finally, we cut and pasted in the figures and spent a frantic weekend making copies of our proposal on a fitfully working astronomy department machine (to save the cost of professional reproduction). My deputy principal investigator, Erick Young, hand-delivered it to NASA Headquarters on November 20, 1983.

I had thrown myself completely into preparing the proposal. SIRTF would give me a chance to do something really noteworthy as an infrared astronomer. Since I was relatively young, I also hoped to finish with SIRTF soon enough to expand on the experience with some future mission.

Seven months later, we were at a conference at the University of California at Santa Cruz when I got a message to call Nancy Boggess at NASA Headquarters. Our proposal had been selected! I was to fly immediately to Washington for a celebration. Santa Cruz conferences are informal. We drove into town to buy a dress shirt, and I borrowed a necktie from a well-dressed French colleague. The next day we gathered at the Cosmopolitan Club. The "Cosmos Club" is an exclusive gathering place for Washington power brokers—so exclusive that it does not admit women. We smuggled Boggess down an alley, into a side door, and up the back stairs so she could join in the festivity for which she deserved more credit than anyone else.

In addition to my proposal for a far infrared imager, Jim Houck's team from Cornell had been picked to build a spectrograph, and Giovanni Fazio's from the Smithsonian Astrophysical Observatory to build a near and mid-infrared imager. Frank Low, of the University of Arizona, had been named as facility scientist (to help oversee the interface between technical and scientific considerations); and Ned Wright and Mike Jura, both from UCLA, were to be the interdisciplinary scientists (to relate the science that would be done with SIRTF to other areas in astronomy). Mike Werner, of Ames Research Center, was the project scientist (charged with overall scientific leadership). Together we constituted the "Science Working Group." In addition to the seven of us, each instrument team included a dozen or so other astronomers; the entire extended SIRTF science team thus had forty-five to fifty members. At the time, it was a significant fraction of the American infrared astronomy community. The project management and engineering team was to be at Ames Research Center at the Moffett Field naval air station just south of San Francisco.

SIRTF would make use of the simple physics of operating a telescope in an environment so cold that it would not drown out the infrared signals we wanted to study with heat—or infrared—radiation of its own. The mission promised huge gains for measurements from wavelengths of 3 microns all the way to 200 microns, a big chunk of the electromagnetic spectrum.[1] Observing in this spectral range would give us a unique perspective on the Universe for four reasons: (1) cool objects, such as planets, emit mostly in the infrared; (2) the infrared is the spectral range in which to study cool states of matter such as solids and molecules; (3) interstellar dust scatters and absorbs visible and ultraviolet light, so in dusty regions even hot objects can best be studied in the infrared; and (4) the cosmic redshift causes the visible light of very distant objects to appear in the infrared.

The potential advantages for infrared astronomy using cold telescopes in space are shown in Figure 1.1. NASA began using such telescopes with the

Infrared Astronomy Satellite (IRAS), a survey mission that was launched in early 1983. The survey was to be followed quickly by another mission with more flexible instruments that could be pointed at sources identified by IRAS and study them in detail, SIRTF.

IRAS had sixty-two detectors operating in four wavelength bands between

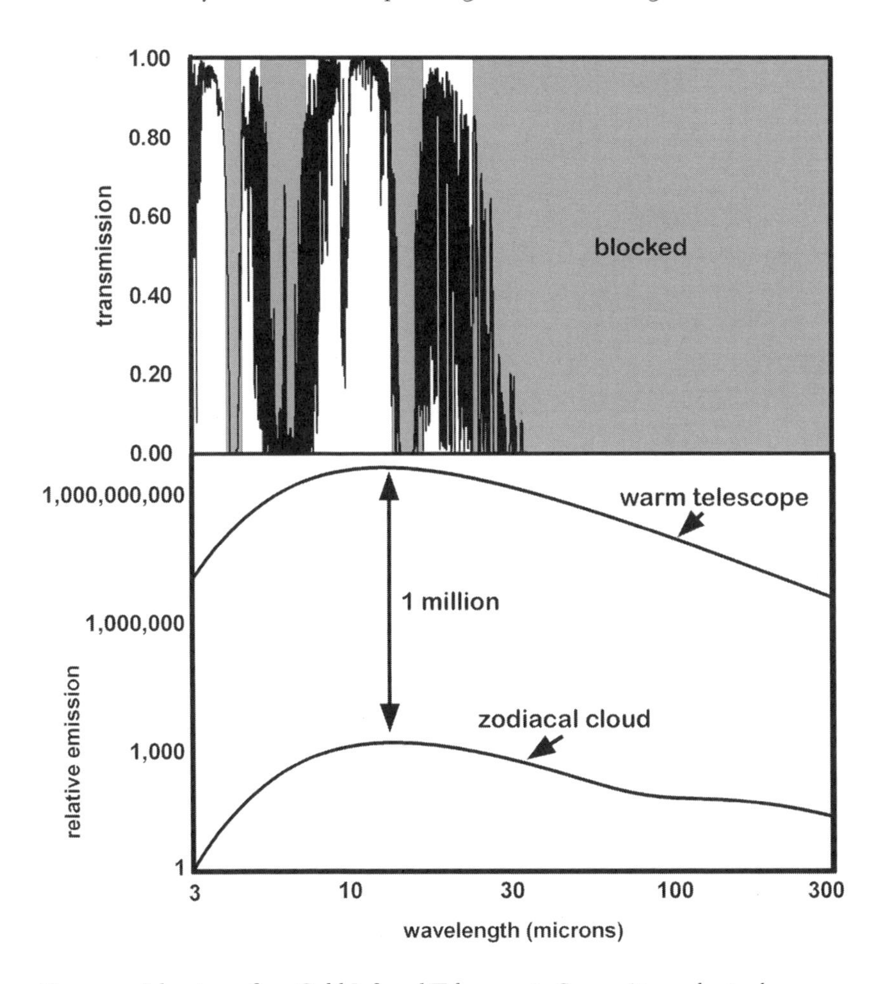

Figure 1.1. Advantages for a Cold Infrared Telescope in Space. Atmospheric absorption makes it impossible for ground-based telescopes to observe between 30 and 300 microns and hampers observations at other wavelengths (upper panel; blocked wavelength ranges are shaded). In space, the entire range is free of interference. The lower panel shows how cooling the telescope drops the interfering foreground radiation from the telescope and atmosphere by a factor of more than one million, to the level of the faint glow of the zodiacal cloud.

12 and 100 microns. The detectors for the longest wavelengths had to operate at a temperature of only 2 degrees Kelvin, or 2K, and to keep from blinding them with thermal emission the telescope needed to be nearly as cold.[2] To survey the entire sky, IRAS had to operate for months in space. Meeting these requirements required the development of a long-life dewar to hold liquid helium.[3] A dewar with an expected life of one year was successfully designed, tested, and delivered along with the rest of the instrumentation by Ball Aerospace. The Netherlands supplied the satellite, and England provided the ground telemetry station.

IRAS was begun on the premise that detector technology developed for military purposes could be adapted to provide the sensory devices. There was no analogy with military sensor systems for the longest-wavelength detector type, however, because there is no military application at these wavelengths. The long wavelength detectors were nevertheless first purchased from military suppliers in the belief that the processes they had developed for other detector types would produce reliable devices. When this belief proved incorrect, additional detectors were built in Jim Houck's lab at Cornell University.

A large classified technology applicable to the shorter-wavelength bands did exist, and it did establish the basis for the IRAS detectors. Even here, however, the potential military contribution had been overestimated. The transistors that read out the detectors were among the worst aspects of the military heritage. When they were destroyed in a fortuitous accident, the decision was made to replace them with a much better performing type. My group had pioneered using these new devices in a comparable application; our approaches were adapted for space by Frank Low's small company.

The development process was far from smooth. In addition to the accident that destroyed the original readouts, large parts of the IRAS detector complement had to be replaced when the original deliveries did not work properly, leaks had to be repaired in the liquid helium dewar, there was an error that degraded the figure of the primary mirror, and the electrical wiring developed various breaks and shorts. The scientists associated with the mission, led by Gerry Neugebauer of Cal Tech, played central roles in dealing with these problems. For a while big issues seemed to emerge almost continuously, and some of the science team members were close to despair. One of them even lost a substantial wager (a case of Scotch whiskey) when the satellite actually worked. A key participant in the final push to launch IRAS summarized the level of desperation by noting with amazement that, unlike most missions, the engineers and scientists had cooperated rather than fighting continuously because both groups realized that failure would be the result if they did not.

IRAS was built in the era when NASA allowed the requirements for a science mission to take priority over budget concerns. IRAS's problems cost NASA far more than the agency had originally allocated, but the extra funds were provided and the mission was built to the original goals—if not beyond them. Fortunately, IRAS had a successful ten-month mission after its launch in early 1983, and it yielded one of the most productive sets of data of any NASA science project.

At that time, Nancy Boggess ran the NASA infrared program under physics and astronomy head Frank Martin. She convinced Martin to start preparing an Announcement of Opportunity to select instrument teams for SIRTF to capitalize on the success of the IRAS mission. The Dutch, who were partners on IRAS, had similar thoughts and were preparing a European proposal for a follow-up mission; Boggess wanted to ensure that the United States was not left behind. Martin left NASA Headquarters at almost the same time IRAS was launched, and Charlie Pellerin took his place. Pellerin and Boggess managed to get the Announcement of Opportunity released, and that is where my part in the SIRTF story begins.

In fact, NASA accepted proposals for a mission concept that had just been made obsolete. At the time the Announcement of Opportunity was written, the NASA upper administration was enamored of a scientist-in-space concept for the space shuttle, and the agency needed to fill the shuttle launch manifest. Using the lack of demonstrated success for long-life helium-cooled telescopes in space as an additional pretext, the Announcement of Opportunity instructed proposers to assume that the telescope would be launched by the shuttle and would operate from the shuttle equipment bay—hence the name Shuttle Infrared Telescope Facility. After two weeks of observation this SIRTF would be brought back to the earth, where it would be refurbished and its helium renewed for another launch. The concept was founded on the illusion, already clearly discredited, that space shuttle launches would provide inexpensive, easy, and frequent access to space. The idea was doomed to failure for other reasons as well. The cloud of contaminants that surrounds the shuttle might glow and compromise the observatory's performance. The contaminants might also slowly freeze onto the cold telescope and coat it with gunk.

IRAS had escaped the shuttle mandate because it required an orbit not accessible by the shuttle. In addition, the Dutch were unwilling to adapt their spacecraft to shuttle regulations. The success of IRAS demonstrated that frequent refilling of the helium dewar was not necessary. As a result, the strictures of the initial Announcement of Opportunity were tempered by an announcement, issued in September, that proposals should anticipate "the possibility of a long duration SIRTF mission." Proposers were instructed to write dual science programs: one using the shuttle and one to describe the science that could be accomplished if the telescope were launched into a free-flying orbit where it could operate for a year-long mission on its own. The advantages for the latter option—increased observation time (by a factor of thirty to one hundred) in a better environment—were so great that it was difficult to give both versions equal treatment.

Although it was now obvious that a shuttle-attached mission would be a mistake, NASA deputy administrator Hans Mark resisted any change. The vision of humans in space enabling great science was too strong to abandon, even after it had been shown to be a mirage. It was NASA policy that every mission that could be launched by the shuttle would be launched that way. For the three years prior to the *Challenger* disaster in January 1986, virtually every science mission was shuttle-launched. Pellerin had to overlook the weaknesses of the shuttle approach to avoid delaying the Announcement of Opportunity: "I knew NASA would eventually do a free flier. But Nancy convinced me to go ahead and get the scientists onboard."

In May 1984, the Ames project team presented a study of the alternative approaches for SIRTF to Burt Edelson, associate administrator of Space Science and Applications, leading to a recommendation in favor of a free-flying satellite. The recommendation was forwarded to NASA administrator James Beggs along with a request to approve the selection of the winning responses to the Announcement of Opportunity. Edelson instructed Ames to study versions of the mission using either an IRAS-type free-flying orbit or a shuttle-accessible one. Only after Hans Mark left NASA on September 1, however, was Pellerin confident he would get final approval for the free-flying version. At Edelson's suggestion, the mission's name was changed to Space Infrared Telescope Facility. Since everyone in NASA refers to missions by their acronyms, the name had the subtle benefit of being old and new at the same time.

Fortunately, we were not fully aware of these political issues. We were too busy during the summer of 1984 with telephone calls, meetings, and negotiations to prepare for building SIRTF to pay much attention. In September, Werner called the Science Working Group together at Ames. We excitedly pondered many issues: choices of orbit, tracking requirements, telescope optical layout, ability to track planetary objects, how to operate SIRTF once it was in orbit, and boosting the mission with Congress. We handed around our first success in the latter arena, a June 1 letter from Jake Garn, chair of the HUD–Independent Agencies subcommittee that oversaw NASA's budget, to Arizona senator Dennis DeConcini: "As per our conversation in Full Committee, I believe the SIRTF program has considerable merit and I have talked with NASA about this issue. I expect that SIRTF will be included in the FY86 budget at a level considerably higher than the $1 to $2 million in FY85. I further expect that NASA will designate it as a New Start in FY87 or FY88. You can be sure I will continue to keep an eye on this program." "New Start" is the term used to indicate NASA and congressional approval for the design and construction of a new mission. Toward the end of the meeting, a group photograph was taken. It records indelibly our relative youthfulness.

2

1985–1989: MARKING TIME

■ The success of IRAS should have given the prospects for SIRTF a huge boost. They did not. What was boosted was the Infrared Space Observatory (ISO), the proposal the Dutch had started in Europe with similar goals. To demonstrate European technology, ISO was to be developed around European detectors and a new European helium dewar. A cooperative U.S.-European mission seemed unlikely. There was substantial pride at NASA in the success of IRAS, and military funding had just produced an exciting breakthrough in the form of the first arrays of infrared detectors with performance at levels that would soon revolutionize infrared astronomy. NASA felt that ISO would be inferior to SIRTF, and the Europeans resented the condescension. Also, over the preceding decade NASA had annoyed the Europeans with a series of unilateral decisions affecting collaborative projects. The competing nationalistic interests negated any attempts at cooperation.

The priorities in American astronomy are set every ten years by the National Academy of Sciences through a study familiarly described as a "decadal survey" and known by the name of the committee chair. The surveys produce a set of priorities that can be taken to funding agencies and Congress for implementation. Even a field as confined as astronomy includes many areas of specialization that compete fiercely for position. To be effective, the final recommendations need to gain acceptance from the whole field. As a result, the survey committees sometimes try to find ways to advance more than one first priority. George Field had chaired the committee that released the 1982 report (Field et al. 1982). Its recommendations addressed what the committee felt were "programs originally proposed for funding in fiscal year 1983 or later." SIRTF was placed among the "projects that were, at the beginning of the survey in 1979, candidates for implementation in fiscal year 1982 or earlier [and] were therefore not considered for inclusion in the Committee's recommendations." Through this hairsplitting logic we were denied the sprinkling of hard-to-come-by National Academy fairy dust that could boost a mission toward approval. The highest *new* priority for the decade was a large X-ray telescope, the Advanced X-ray Astrophysics Facility (AXAF).

Normally, it would have been plausible to build both SIRTF and AXAF within the same decade. NASA planners like to represent year-to-year funding profiles in charts that look like geological strata, each layer representing an individual project and the total stack showing the total funding in the program. Unlike the geological analogy, a given layer is supposed to thin down as the project is completed, making room for another layer within the stack—a new project. However, the already approved Space Science missions were all demanding more money, and the thinning process was taking longer than anticipated. In particular, the Hubble Space Telescope (HST) was consuming a third of the physics and astronomy budget, leaving little room for another large mission after its bills had been paid. For a while, the growth in the estimated cost to complete Hubble outstripped the rate at which money was spent: in February 1980 the total bill was pegged at $530 million, but by April 1984 this estimate had ballooned to $1.2 billion. After the *Challenger* disaster in 1986 grounded all NASA launches, Hubble had to be maintained (and first completed—it was not going to be ready to launch on its original schedule), and as a result, its fat stratum in the NASA budget charts continued through the entire decade.

Originally, space science had been pursued by a small number of risk-taking pioneers, and competition for missions had been mild. Sometimes, NASA Headquarters administrators had simply picked experiments themselves (Roman 2000). With time, the technology improved and success became more probable. Funding became abundant as a by-product of the Cold War and the desire to obliterate the memory of the USSR's Sputnik beating the United States into space. The mid-1970s had seen a sequence of ambitious NASA Space Science missions: Vikings to Mars, Voyagers to the outer planets. By the mid-1980s, however, the good times had come to an abrupt end. In 1986, the NASA Space and Earth Science Advisory Committee concluded that the Space Science program was "facing grave difficulties" leading to "a growing sense of unease and frustration over the program's diminishing pace." The committee noted that more and more missions were being identified as candidates for New Starts at a time when prospects for New Starts were becoming uncertain. As a result, "the competition among prospective missions had escalated to a counterproductive level" (Space and Earth Science Committee 1986).

In November 1981, Space Science had been merged with Space Applications into the Office of Space Science and Applications (OSSA). In February 1982, Burt Edelson had become the associate administrator for this combined office. Edelson did not try to prioritize science missions; he left it to the different science areas to develop missions and to vie with each other for approval. Every year he held a retreat in which the priorities for new missions within OSSA were fought out from the very beginning, without consideration for previous rankings. By its nature, such an allocation of resources tended toward the near view, since small, short-term endeavors could be fitted more easily into modest year-to-year opportunities. Long-term major endeavors like AXAF and SIRTF were at a huge disadvantage.

In this difficult environment, physics and astronomy head Pellerin needed to promote *both* AXAF and SIRTF, and to do it before the bills had been paid for the Hubble Telescope. It was going to require a sophisticated selling approach. He called together a group of scientists to develop ways to emphasize how combining studies across the entire electromagnetic spectrum, linking the already approved Hubble Space Telescope and Gamma Ray Observatory with the unapproved AXAF and SIRTF, would lead to major breakthroughs. To keep the work approachable, he told the group, "We're going to make a coloring book to lay out the entire program so everyone would see the need for all of it," and he provided paper and crayons.

To follow up, he established a slightly more formal "Astrophysics Council." We were given the crayon drawings of the preceding group and worked under the tutelage of a professional artist to improve them for what amounted to a sales pamphlet. The initial idea was to promote the suite of missions as a "New Era for Astronomy," but after a few meetings the more evocative term "Great Observatories" was suggested by George Field. Each of the Great Observatories explored a separate part of the electromagnetic spectrum. The drawings showed how they were complementary and together would advance our understanding broadly. When the pamphlet emerged, its brightly colored illustrations had an abstract air that seemed to place the science missions and their objectives in an ideal world, somewhere between reality and an intellectual form of science fiction.

The NASA orthodoxy in 1984 had moved beyond the shuttle and held that fledgling missions could use the space station, thus providing long-duration missions but still relying heavily on the manned program. In our view, the station had many of the same disadvantages as the shuttle. In attempts to convince scientists that the station was a useful concept, NASA added conceptual outposts in the form of free-flying satellites. At one point Edelson solemnly promised me that he would get SIRTF a New Start if I would convince the SIRTF community to put the telescope on one of these station outposts. The concept was a little too weird to take seriously, and both Edelson and the outposts disappeared before any convincing became possible.

Despite the attractive Great Observatories brochure and its vision of peaceful coexistence for all four missions, NASA could promote only one observatory at a time. AXAF had a big head start: it had received the Field report's recommendation, and mission concept ("Phase A") studies for it had been under way since 1977. The delay to late 1984 in declaring the obvious—that SIRTF should be a free flyer—had postponed any meaningful Phase A work on it at least into 1985. Pellerin gravitated quickly to AXAF. We protested that the Field report's assumption that we were already approved clearly implied that we should not be shoved aside. Our requests to have the undefined relative priorities clarified were ignored. Relations were strained: at our May 1985 meeting, Pellerin scolded us for "threatening to disrupt the Field report." In mid-1985, Pellerin explained his plans to the Astrophysics Council for the next budget cycle: getting AXAF under way, a lot of effort on lesser endeavors, but not a word about SIRTF. I blurted out, "But what about SIRTF?" "You infrared astronomers are always yapping at me," he snapped back. I was promptly replaced on the council.

Despite getting Pellerin's support, AXAF did not get started until 1989, after its own series of near-death experiences (Tucker and Tucker 2001). It was launched in 1999 and renamed Chandra after the brilliant astrophysical theoretician S. Chandrasekhar.

The SIRTF Science Working Group certainly felt the "unease and frustration" reported by the Space and Earth Science Advisory Committee. We were restive, and wrangling often marred our periodic meetings. Nonetheless, SIRTF continued to be put through a series of transformations to keep it in line with whatever orthodoxy was current for NASA missions. We never lost hope that we could reinvent it into a mission that would be approved. For the most part these efforts appeared to lead nowhere, although they did perversely produce a skill in mission and instrument redesign that proved critical in adapting to radically changing realities within NASA.

Although we now knew that SIRTF was to be "free flying," no one was quite sure what that meant. An initial polar orbit concept was based loosely on IRAS, but it was really hardly more than an artist's sketch. In any case, SIRTF would be too heavy for this plan. Quickly, the designs returned to orbits accessible by the shuttle. Through the first half of 1985, SIRTF concepts looked as if the old shuttle-attached version had been cut loose where it had fastened to the shuttle bay, with a spacecraft bolted on to cover the wound. This contraption was to operate from a "low-earth orbit" directly accessible by the shuttle. The problem with such orbits is that the earth fills nearly half the view of space. For most astronomy missions, such as the Hubble Telescope, the earth is an opaque nuisance that reduces efficiency by blocking the view of space in a given direction for half of every orbit. The situation would be far worse for a helium-cooled telescope. The earth is a huge source of heat radiation that would warm the telescope and boil away its precious helium coolant. The sun is another enemy, and for the part of its orbit where SIRTF passed between the sun and the earth, the pointing system would have had to thread a sinuous path between the two to avoid disaster. Observational efficiency was going to be poor, and a small error along that path could terminate the mission—quickly. Nonetheless, with little understanding at the time of the error-prone nature of space operations, I listened to these proposals gullibly, hoping they would get us going.

By the end of 1985, the problems that would arise from the low-earth orbit had been addressed by appeal to Flash Gordon—or so it seemed. The telescope would still be launched by the shuttle, but nestled with it in the equipment bay would be a rocket that looked like an oversized elephant stand, the Orbital Maneuvering Vehicle (OMV). NASA was planning to build the OMV to overcome the limitations in orbits accessible directly by the shuttle. The OMV would boost satellites into different orbits, then return to the shuttle bay. In this new conception, the OMV would boost SIRTF from the shuttle orbit, at an altitude of about 250 kilometers, into a higher orbit near 900 kilometers. Periodically, when the helium coolant was about to run out, the shuttle would be launched with the OMV and a load of liquid helium. The OMV would be dispatched to fetch SIRTF and bring it down to the shuttle, and astronauts wearing heavy gloves and floating in awkward spacesuits would transfer a new load of helium—a tricky operation even in a laboratory attached securely to the surface of the earth. Before being strapped into the shuttle for return to the earth the OMV would carry the telescope back to its 900-kilometer orbit.

Although this approach reduced some of the extreme disadvantages of the shuttle-accessible SIRTF, it still left the telescope closer to the earth than was desirable to keep it cold. This time it was apparent to me that the tradeoff in complexity, expense, and risk was both ludicrous and frightening. I suppressed these thoughts, since there was no alternative in sight. Although the OMV was enthusiastically promoted, it was eventually canceled.

We brainstormed. In some dark hour when it appeared that we would not get started within our lifetimes, someone made a radical suggestion. Why not convince Ball Aerospace to build a SIRTF as close to IRAS in design as possible, on a fixed-cost basis so NASA would not have to contend with budget overruns? That would eliminate many of the costs of new designs and their qualification for spaceflight, and a low-cost mission could get under way quickly. We all took a blood oath to keep this idea secret. We did not want word of it to reach NASA Headquarters, lest they impose a much cheaper and less capable mission on us whether we liked it or not.

Our situation was increasingly frustrating. Within the United States, we had been shoved aside by AXAF. In 1986, the European Space Agency (ESA) proceeded with development of ISO. Negotiations between ESA and NASA to coordinate the two missions became increasingly rancorous. Pellerin's deputy, George Newton, forbade us from making any official contact, adding sarcastically that as private citizens we were of course "welcome" to get in touch at our own expense. Our status in the Field report—"assumed to be approved" for construction in the 1980s—had a bitterly ironic feel.

In April 1987, Lennard Fisk became the associate administrator for Space Science. Fisk quickly brought order to the program by establishing a "strategic plan." Although astronomy had been carrying out strategic planning on a regular basis in the form of the National Academy decadal surveys, such a systematic approach was uncommon in NASA. Under Fisk, potential new Space Science projects of different scales were ranked on different lists for New Starts. There were explicit rules determining from which list a highly ranked project would be selected. For example, a modest new funding opportunity would lead to selection of the highest item on the list of modest projects. The different science disciplines knew where their missions stood and were encouraged to support the whole program. Fisk made it clear that attempts to leapfrog up the priority queues by zealous advocacy would not be tolerated. Strategic plans proved so effective that they have remained a centerpiece of the program ever since.

At about the same time, Fred Gillett began a two-year sabbatical as a visiting senior scientist at NASA Headquarters. Fred had been a member of the IRAS science team and had a uniquely quiet and thoughtful air of authority. He carefully selected members of an infrared advisory committee, chaired by Chas Beichman, that put together a strategy for infrared astronomy.

The shuttles were cleared for further launches in late 1988, starting with *Discovery*. *Magellan* was launched in May 1989, and the next four years saw ten or so shuttle launches per year. It was, in Fisk's words, "like the Fourth of July. . . . We had everything going up. It was a wonderful time to be there" (Fisk 2000). Missions had been building up for years, and when the shuttles started flying again, space science really took off.

The Augustine report had just projected an era of strong growth for science within NASA. The committee recommended that a new division be created for Earth Observing Systems (EOS) so that this huge enterprise could proceed without affecting the funding for other areas; basically it was a proposal to increase the science share of the NASA budget dramatically (Augustine et al. 1990). Fisk wrote into his plan the "expectation" that NASA would start one moderate or major mission per year, a dramatic reversal from the dismal rate of mission approval for the previous five years. Since there were four science divisions under him, his rules implied one start per division per four years. If he could get AXAF started, we could hope to proceed four years later.

Fisk later noted that 1989, 1990, and 1991 were "banner years. . . . We sold AXAF, we sold CRAF/Cassini, and then we went for the whole enchilada and sold EOS, which has a $17 billion price tag on it. The budget was literally wonderful at that point. . . . We really had a phenomenal program" (Tucker and Tucker 2001).

Although SIRTF was still alive, neither the overly complex OMV-launched version nor the inadequate fixed-cost one was a viable mission concept. A new idea was needed, and quickly, because a mission without a viable baseline was more likely to be put forward for elimination than for a New Start. Boosting SIRTF to a still higher orbit was the right direction to go. The NASA–Ames-based engineering team gravitated toward launch by the largest possible unmanned rocket, the Titan IV, into an orbit very high above the earth. Studies showed that 100,000 kilometers, 40 percent of the way to the moon, would be about right. This orbit would put the telescope above the trapped radiation zones around the earth—a region where high-energy protons would strike the sensitive infrared detectors and interfere with their operation—but was still close enough to the earth to make it easy to send the data back. We worked feverishly on this new approach. For the first time, a version of the mission emerged that had no serious technical flaws. However, it departed so far from the previous shuttle-based concepts that we needed to obtain approval from Fisk. When the day came, we assembled at NASA Headquarters in Washington and rehearsed all morning, then made the presentation that afternoon. It was carefully choreographed to combine sage advice from elder members of the group, details of the concept from the engineers, and enthusiasm from the rest of us sitting as witnesses. At the end, Fisk said the five important words: "Let's do it that way." Then he got up and left the room.

With AXAF getting a New Start, we should have been poised to move ahead. We were ready technically and were moving upward within the NASA priorities. We would also need approval from Congress, though, and we were no further along with them than when we were first selected in 1984. There are a number of reasons for our dismal political performance. Perhaps most important, we had difficulties mustering the level of commitment necessary to sell a mission. Although infrared astronomy can be done much better from space, with the newly available detector arrays ground-based infrared astronomy had begun to flower. X-ray astronomy, on the other hand, can be done only in space. Because they had already promoted a series of smaller missions, and their professional livelihood depended on success, X-ray astronomers had been much more energetic and effective in explaining the case for AXAF to Congress than we were with SIRTF. This tendency was enhanced because AXAF was primarily an East Coast mission, making it convenient for its advocates to stop by in Washington to promote it.

There may be a deeper reason for our bumbling in SIRTF advocacy. Capitol Hill was an alien culture. Scientists are used to operating in a well-defined environment with clear rules for success. Good grades in technical courses in college lead to admission to graduate school, where an insightful Ph.D. dissertation leads to a postdoctoral position, which with good output leads to a faculty position, where high-quality research yields tenure. Although the "scientific method" is overly simplified in most textbooks, the scientific enterprise does have clear standards for recognizing good research. Scientists are generally selected by their success in this relatively deterministic and rational environment. In contrast, politics is a morass. Everyone knows of meritorious people who worked all their lives toward political goals that were never achieved because of situations over which they had little control. Scientists are often uncomfortable contending with the unpredictability of political advocacy, and only a few of them do it with the zest and finesse that it requires.

As in politics, there is a principle of fair play in science that requires scientists who write proposals that are not selected for a mission to close ranks and support the mission politically. However, as time went on and we made no progress toward realizing our goal, other infrared projects began to view us as an obstacle. The core SIRTF support had shrunk virtually to the seven members of the Science Working Group. We needed outside help to gather the political momentum required to start our mission.

3

1990: A NEW DECADE BRINGS NEW HOPE

■ A full decade had passed, and it was time for the National Academy of Sciences to conduct another decadal survey. We had a new chance to climb to the top of the astronomical priorities and to reap the benefits with Congress. The survey committee was to be led by John Bahcall of the Institute for Advanced Study. Afterward, he described the priority-setting process:

> What are the most important aspects of the universe to explore? What are the best ways to make discoveries in astronomy and astrophysics? These are tough questions because researchers have many different approaches and it is usually not clear, until the most interesting problems are solved, which method will yield the most important results.
>
> We are well into an era of limited research budgets, however, and choices have to be made. We astronomers have recognized that if we do not set our own priorities, then funding agencies and congressional officials will do it for us, and will do it less well. . . .
>
> . . . Much of the difficult work of the survey was done within . . . panels. . . . [They] began their technical work with essays submitted by individual panel members on what they identified as the most important issues or projects. . . . A core group within each panel wrote the

> initial draft of the report, which was then iterated within the panel. . . . The most intense discussions in the first nine months of the survey occurred within the panels.
>
> The survey was organized so that every astronomer who had something to say had an opportunity to be heard. Open discussions were held in conjunction with meetings of the American Astronomical Society (AAS) and at several other professional society meetings. In January 1990, at the Washington, D.C., meeting of the AAS, nearly 1,000 astronomers participated in open sessions that involved all 15 of the panels.
>
> The list of priorities was established by a gradual process that was much easier than any of us anticipated. The committee voted on straw ballots on three occasions during our regularly scheduled meetings, using as background material the preliminary reports of the advisory panels. The straw ballots focused the discussion on projects that were most likely to be considered important in the final deliberations. As preparation for the final ballot, the committee heard advocacy presentations from the panel chairs. The chairs also participated in discussions of the relative merits of all the initiatives, although the final recommendations were formulated by the survey committee in executive session. (Bahcall 1999)

Bahcall's committee was established in May 1989. Fred Gillett's advisory committee released its strategy for infrared astronomy the next month, timed to have as much influence as possible. Chas Beichman had taken a sabbatical from overseeing the reduction of IRAS data to assist Bahcall in pulling the report together. Gillett chaired the infrared advisory panel, and Jim Houck sat on the Bahcall committee itself. Unlike previous surveys, infrared astronomers fresh from a highly successful space mission were thoroughly involved in the process. They had great credibility arguing that new infrared detector technologies operating in another cold telescope in space would bring even greater advances.

The new study concluded: "The highest priority for a major new program in space-based astronomy is the Space Infrared Telescope Facility (SIRTF), a 0.9-m cooled telescope in a spacecraft to be launched by a Titan IV–Centaur into high earth orbit. . . . Across the wavelength region from 3 to 200 microns, SIRTF will be up to a thousand times more sensitive than other space- or ground-based telescopes, and over a million times faster than other instruments for surveying, mapping, or obtaining spectra across large, complex regions" (Bahcall et al. 1991). Beichman recalls that SIRTF was the least contentious of the recommendations; it was obvious that it had huge discovery potential, and astronomers had just had a delicious taste of space infrared astronomy in the IRAS data.

We were certain that the first results from HST would provide a huge boost for astronomy. The combination of the top recommendation from the Bahcall committee and a high level of overall enthusiasm for astronomy would surely generate enough support to get SIRTF under way. All we had to do was wait for HST to get into orbit and return some stunning images.

The prospects for our mission were good enough that NASA began to consider seriously whether Ames Research Center was the right place to build SIRTF. Ames's role within NASA centered on experimental aircraft; huge wind tunnels and airplane and dirigible hangers dominate the landscape within the center. In 1969, when Frank Low convinced NASA to explore the far infrared sky with a small telescope mounted in the side of a Lear Jet, Ames had both the facilities and the pilots he needed, and hence the center had become the focus for airborne infrared astronomy. IRAS also had been started at Ames. When the project got into technical trouble, however, it was transferred to the Jet Propulsion Laboratory (JPL) in Pasadena, leaving behind a cloud of resentment at Ames. Ames retained its interest in infrared astronomy through the Kuiper Airborne Observatory. The successor to the Lear Jet, Kuiper had a 36-inch telescope looking out the side of a C-141 military cargo airplane.

The team of engineers and managers at Ames had devoted many years to SIRTF. When he met with us, the Ames director, Bill Ballhaus, repeatedly told us that we were his second or third priority, behind only mandatory items like emergency wind tunnel repairs. In more public presentations, however, SIRTF tended to drop off the bottom of Ballhaus's standard list of the ten highest priorities at the center. Ames was promoting a new airborne telescope with a 2.5-meter aperture to be mounted in a Boeing 747. There was a risk that the small infrared group would become overextended if both projects went forward together. Ames was basically a research center, not a center for spaceflight. It was unclear whether the management fully appreciated the intricacies of developing a major flight mission. So long as SIRTF had no real chance, these shortcomings were irrelevant. However, when Pellerin saw one of Ballhaus's priority lists with SIRTF at number fourteen, it triggered a chain of reactions. Pellerin held a competition to which Ames, Goddard, and JPL all submitted proposals. Although Ames's proposal was judged to be excellent, it was decided in late 1989 to transfer SIRTF to JPL, to the horror of those at Ames who had seen the same thing happen a decade earlier with IRAS.

The Hubble Space Telescope was finally launched on April 25, 1990. Everything went well until it was pointed at a star. Weeks of agonizingly slow adjustments and tests brought no improvement in the fuzzy images. HST was to educate the public not on astronomy but on "spherical aberration," the optical fault that dominated its images and made Hubble useless for most of its intended science applications. NASA likes to operate in an aura of superhuman perfection, but the HST problem made it the butt of editorial cartoons and television talk show monologues across the nation.

Meanwhile, JPL pressed ahead to establish the infrastructure for SIRTF. A senior manager, Dick Spehalski, who had just finished managing the Galileo mission to Jupiter, was appointed to head the project office. He assembled a team that took over the Titan-launched high-orbit SIRTF and began to fill in the details and refine the cost estimates. Spehalski, gruff and forceful, was experienced in the JPL system. He had a strong sense of order, and his past experience was with planetary missions in which the science teams provided specifications for the instruments but then kept out of the way while the managers and engineers built them.

SIRTF had always been under the protection of the scientists who conceived it, many of whom were renowned for their technical expertise and had played important roles in IRAS. The recent move from Ames included a few brochures and documents, but only the project scientist, Mike Werner, and one other scientist transferred to JPL. At Ames, the mission had been strongly influenced by the Science Working Group, which, despite its name, spent virtually all of its meetings discussing the technical optimization of the mission rather than the science it might accomplish. The scientists naturally felt they were carrying the collective memory and soul of the project from Ames to JPL. They expected a project culture at the new center that would respond to their ideas about the mission and would work closely with them to implement those ideas.

Because of the wide difference in their perspectives, the interface between Spehalski and the scientific leadership of SIRTF was uneasy at best. Even a retreat hosted by Giovanni Fazio in Boston, as far from the JPL environment as possible, was unable to generate a collegial relationship. Nonetheless, we had a highly recommended project to carry out, and we were motivated to work together. We concentrated on key technical developments and on advancing the designs of the instruments, telescope, and spacecraft sufficiently to be able to generate realistic pricing.

JPL took pains to educate us in their conservative and rigid rules for development of a project on the scale of SIRTF. The instrument managers in the project office expressed dread on our behalf for the JPL standards soon to be imposed on us. The office in charge of electronics parts impressed upon us that we had to use the very highest quality (by JPL guidelines) parts available, regardless of any schedule slips or cost overruns that might result. There was to be no compromise in a mission that would use every aspect of the NASA bureaucracy to produce the most reliable product possible. Everything was to be done by that particular very large book.

Contracts were drawn up to get the instruments started toward "Phase B" studies, defined by NASA as the level at which engineers would set a concept firmly and allow it to be priced accurately. After a period of long-distance exchange on the details of the contract for my instrument, a team from JPL came to close out negotiations on January 17, 1991. Near the end, our secretary peeked through the door and told us that U.S. warplanes were bombing Baghdad. We concluded the negotiations listlessly. Everyone in the room had a foreboding that the demands of a war would change the terms far beyond the details we had been arguing about.

We also took steps to improve our salesmanship. My wife, Marcia, had been appointed our public outreach officer in December 1990. On January 17, she was in Washington making her first round of congressional visits. As she was discussing final strategies at NASA Headquarters, a letter arrived from Senator Barbara Mikulski's chief of staff, Kevin Kelly, instructing NASA to back down on SIRTF. Mikulski, who chaired the relevant appropriations committee, had decided that NASA's funding could support no new Phase B studies.

Marcia had a lonely, miserable tour of Washington.

The HST failure, the war, and the Mikulski-Kelly letter were not all. In April, the Galileo spacecraft executed the commands to open its high-gain radio antenna, essential to send data from Jupiter back to Earth. The telemetry showed that the motors had stalled and the antenna had only partially opened—into a distorted shape that made it useless. The mission would have to be conducted on the backup low-gain antenna, which produced signals about ten thousand times fainter.

Despite our failure to get into Phase B and the gathering storm clouds over large NASA science missions, Spehalski directed us to push on and get cost estimates for all the elements of SIRTF. As the estimates came in, it became clear that they would be well above the $1.3 billion price tag adopted by the Bahcall report. To help deal with this issue, Spehalski assigned a team to make independent cost estimates for each instrument. These estimates could be predicted to exceed substantially those generated by the teams themselves, particularly since they were conducted in sufficient isolation to ensure a number of misunderstandings. The teams were to be confronted with these estimates and forced to make significant reductions in the instrument capabilities, and thus in their cost.

The JPL team used computer-based cost models that included parameters such as the estimated weight of the instrument, the number of optical elements and moving parts, general tolerances, and so forth. The estimators could take these parameters directly from preliminary designs. However, the models also made adjustments for technical risk, how far the design departed from established practice, the experience with similar devices in previous NASA instruments, and the level and quality of the prior development work. All of the inputs beyond weight and number of parts and mechanisms were subjective and required detailed understanding of the instrument and its state of development—subjects about which the modelers showed little interest. If these inputs were offered merely as the opinions of the cost estimators, they would have little credibility. If an inscrutable computer model is employed, however, the results are hard to challenge. Our estimator at Ball Aerospace threw in a broad grin for free and summarized the process as "garbage in, gospel out." The JPL estimators did not display the same self-deprecating humor.

We were the first instrument team called to JPL to confront the numbers. The estimates for all three SIRTF instruments added up to nearly $500 million. The estimators thought our instrument would take about $160 million to complete, whereas our own estimate placed it just above $100 million. The difference is easily ascribed to the subjective factors just described, and we entered into a detailed debate about the JPL estimators' assumptions. The estimators were ignorant of the progress we had made on technical aspects of the instrument and hence had assigned inappropriately large risk factors. Eventually, they agreed that it was possible that our estimate was correct. Since traditionally $100 million was an appropriate share of a $1.3 billion mission, there was an uneasy silence while the cost estimators looked to Spehalski to indicate the next step.

Before he could respond, he was called out of the room. He returned about five minutes later and sat for the rest of the afternoon without making a comment while we wrangled over details of the two cost estimates. As we finished, Spehalski asked me to stop by his office. "That was a telephone call from Washington," he said. "SIRTF has been canceled." Someone lobbying for Lockheed on behalf of the Orbiting Solar Laboratory, a mission that had been postponed along with ours, had resisted the cancellation of their Phase B study too vigorously. Their protests had produced a whiplash in which, for emphasis, NASA was directed not to spend anything at all on the missions that had just had their Phase B studies delayed.

Why had SIRTF been treated so harshly? Congress has the responsibility to keep a lid on NASA expenditures, letting major missions get started at a measured pace to stay within national priorities. Starting a new mission involves a commitment to provide funding for a decade or more, first to pay for the design and construction, then for the launch and operations in space. NASA tries to keep a stable of missions under active development awaiting congressional approval, counting on building constituencies that will apply pressure to get official starts. Congress tries to resist this strategy and to preserve flexibility to adjust the NASA budget downward, a difficult task in the face of these existing decade-long commitments.

Congress, in other words, was suffering from a huge case of buyer's remorse. Space Science director Fisk's vision and organization had worked too well. He had sold an underlying assumption that there would be 25 percent annual increases in the Space Sciences budget. In fact, NASA's budget position weakened as the economy slowed down in the recession of 1990–1991. In addition, CRAF/Cassini, originally promoted as a dual mission to a comet and to Saturn for hardly more than the cost of a single one, announced a projected budget increase from $1.6 to $2.1 billion, and AXAF had grown by a similar amount (GAO 1994).

Other recipients of discretionary spending that had to share funds with NASA (like veterans' benefits, housing, and education) were not going to make room for NASA spending increases in the face of a tightening overall budget. The problems in the HST optics and Galileo antenna dramatized the risks associated with expensive science missions. The space station was starting to eat up a huge part of NASA's budget. An attempt to rally scientists and others to oppose the station in Congress led to a dramatic demonstration of the power of the professional aerospace lobby over the amateurish scientific one—the station was easily victorious.

Even the newly approved missions were under severe threat. Within NASA, there were hasty efforts to economize by trimming missions in early development. AXAF was redesigned to leave off one instrument, reduce the number of imaging mirrors, and change the orbit to decrease operational costs. The congressional commitment to CRAF (Comet Rendezvous–Asteroid Flyby) was withdrawn to contain cost growth. As Fisk described it: "There was a sea change in Space Science: we had to modify AXAF, we lost CRAF, EOS took a real beating."

New missions were to be put off until they fit more easily into the funding available, perhaps forever. SIRTF was ramped down rapidly at JPL. Spehalski himself transferred to managing the Saturn mission, Cassini. However, even dead NASA projects cannot escape the ubiquitous reviews. Fifteen JPL personnel—all of whom were to be laid off in two weeks—were flown to Ball for a previously scheduled review of our instrument. We could afford to bring in the Ball engineers only for their specific presentations, after which they were sent back to their new responsibilities elsewhere within the company. With the exception of a few who slept through most of the proceedings, the reviewers appeared not to be aware that they were about to be reassigned. We were queried severely, with strong emphasis on applying JPL quality assurance standards to our nonexistent instrument. After the review, we had a farewell dinner at a local restaurant for the Ball personnel who had worked with us, expecting never to see them again.

Fortunately, neither NASA nor the SIRTF scientists gave up. Fisk felt a strong commitment to the mission because as one of the four Great Observatories it was a central NASA program. He set out to keep the project alive, if in hibernation. He pointed out to Kevin Kelly that most of NASA's funds to develop infrared detectors were concentrated in the SIRTF teams. Wiping out SIRTF amounted to a deep cut in the funding for detector development that would cripple NASA's entire future in infrared research. Kelly relented just enough to permit NASA to fund the instrument teams to continue with this part of the program. The industrial instrument design efforts were still embargoed. For more than a year, the acronym "SIRTF" was banned in the halls of Congress.

As soon as possible, we renewed our efforts to persuade Congress to fund SIRTF. Projects that require government funds are usually approved in one of two ways. In the first, a congressman may push through an appropriation to bring business to his district. However, no science mission, and certainly no mission on a scale plausible for SIRTF, could provide sufficient return to a single congressional district to co-opt its congressional representatives. Not only are the budgets of science missions relatively small by this measure, but the spending is spread over many states (six in the case of SIRTF), and their larger constituency (scientists who will use data from the mission) is more thinly spread yet. The second way—some tycoon or large company contributing heavily to political campaigns and gaining access to people with power—certainly does *not* apply to scientists.

Although we expected the congressmen and senators from the districts and states that would have some involvement with SIRTF to be supportive, we knew that additional support had to be found. Science projects are usually not Democratic or Republican. There are always a few Democrats who wonder why we are spending the money on expensive science rather than social programs and a few Republicans who think NASA should be broken up and privatized, but supporters of NASA science are scattered all over the vast center ground occupied by both parties. Unfortunately, there is no directory of such people; you have to find them yourself. Word of mouth is the starting point, but it seldom suffices. The next step is to make a list of key committee members and people on standing committees and start visiting them. And that is what we did.

A visit to a congressman's office is a flattering experience. The offices are tastefully decorated with trappings both of power and of the home district. A receptionist makes sure you are seated comfortably and offers you a cup of coffee or tea. At the time of your appointment you are ushered into an interview room, where a very polite and sympathetic (and usually young) staff person listens carefully to your pitch. At the end, she thanks you for the interesting information and promises to bring it to the attention of the congressman. After a few such experiences, you realize that they are formulaic. Every member of Congress has an obligation to meet with constituents—almost always people wanting something—and every member hires junior staff to talk with these people and to promise to bring the problem to the attention of the congressman. And undoubtedly the promise is honored—in the form of a very short paragraph in a daily diary that the congressman scans perfunctorily before going home.

Occasionally, though, you make contact with someone with a real interest in the project. It might not arise from a scientific motive, but that does not really matter. In the case of SIRTF, we found a lot of enthusiasm in a key congressman's office from a staff member who was interested in the economic potential of mining asteroids and was duly impressed at how readily our mission could detect them. Another person was strongly influenced by the social structure of our effort at the University of Arizona—the mixture of backgrounds, races, and sexes who were all working on the project, and the diversity of interests, including undergraduate and precollegiate education.

The most successful NASA science mission in congressional relations has to be "Gravity Probe B." Although he is based on the West Coast at Stanford, the principal investigator, Francis Everett, seemed to have time to attend every activity in Washington that might affect his project. Everett looks like Albert Einstein's younger brother, so his appearances were hard to miss. More important, he adopted a strategy of maintaining contacts for the long haul, and from both political parties. Everett was so successful that his project survived a number of attempts to cancel it from within NASA as well as lukewarm recommendations from outside advisory committees, any one of which would usually be sufficient to eliminate a mission.

In general, scientists have liberal political leanings that make them more at ease with Democrats. The long period of Democratic control of Congress up to 1994 made it easy to assume that Republicans were not an essential part of the round of Washington visits. However, after talking to Everett, Marcia decided to give equal attention to the shadow government of potential Republican chairs and their staff members for the relevant committees. She continued to make the rounds in Washington, promoting SIRTF and seeking out people who would help. There was a surprisingly large number; astronomy has enormous appeal to the public—and evidently to public representatives as well.

Finding friends scattered through Congress was not going to be enough to get SIRTF started. Fisk needed to update his strategic plan to reflect the new reality. A new plan could also activate the space science community and continue the pressure on Congress for growth in his budget.

Reaching a consensus in the whole space science community would be far more difficult than the consensus building necessary for a decadal report on astronomy. The various fields of space science—space physics, space biology, microgravity, astronomy, planetary studies—cover a broad range of disciplines. Successful researchers need a high degree of focus and enthusiasm for their specialty. Few have the same level of enthusiasm for fields not directly related to their own. This tendency is reinforced because priority-setting groups are selected in a way that makes each member feel responsible for representing his or her field. Furthermore, some of the fields had experienced the same kind of difficulties that had frustrated SIRTF, making their advocates impatient and disinclined to compromise.

Nonetheless, science priorities had to be set. The usual approach is for the administrators to lure representatives of all the fields into a room, lock the door, and not let anyone out until some kind of agreement is reached. It can be taken for granted in such a negotiation that most scientists will vote their particular field at top priority. The key to success is to garner the majority of second-place votes from other fields.

Although the procedure can be brutal, an attempt is made to provide as attractive and comfortable a room as possible. Thus, a high-level meeting involving representatives of all the science disciplines (with Marcia for astrophysics and particularly for SIRTF) was held at Woods Hole, at the base of Cape Cod, in early August 1991. After four days of stormy negotiation (including a heated discussion of one another's ancestry by NASA Astrophysics head Pellerin and a notoriously aggressive solar physicist), the delegates emerged with the outline of the new plan. Although the other areas spread their second-place votes widely, astronomers and planetary scientists had turned out to have enough overlap in goals that each had supported the other's priorities as a backup to their own. Both fields had emerged well placed in the new plan. SIRTF was the top-priority "Flagship" mission (the committee had adopted this term to avoid the by now politically incorrect "large" terminology). Other fields did not do so well. The Orbiting Solar Laboratory was demoted and quickly disappeared.

4

1992: FASTER, BETTER, CHEAPER

■ After more than a year of detector-only development, NASA managed to get approval to reconsider SIRTF. By then, however, NASA was a different agency, one struggling to redefine itself and with an imperative to control costs and simplify projects. Mission requirements were now to adapt to budgets, rather than the other way around. A new management style was also in vogue.

Management styles have a complex history within NASA. At the inception of the Apollo program, NASA encountered nearly disastrous problems in its approach to managing large projects. These issues had been addressed by importing a systems management structure from the military. Under this approach, different aspects of a project were supervised carefully to be sure that they were adhering to a set of consistent projectwide objectives. Performance was monitored in person and day to day by project representatives, and additionally by periodic reviews. There were complex rules for purchase and qualification of parts. Complicated analyses searched for failure possibilities. Systems management contributed critically to the success of the Apollo program. As time passed, however, the rigidity and complexity of these procedures grew and they contributed to substantial delays and huge cost growth. A General Accounting Office study showed that missions typically had nearly doubled in cost from approval to launch (Roy 1998). We had been setting out to build SIRTF exactly according to these rules when the project had been canceled.

The White House Space Council, chaired by Vice President Dan Quayle, felt that the conservatism in the agency threatened President George Bush's "Space Exploration Initiative," an ambitious proposal made in late 1989 to send astronauts to Mars. Again, they looked to the military—this time to the Strategic Defense Initiative Organization (SDIO), or "Star Wars"—for more efficient management approaches. When President Ronald Reagan had proposed SDIO, it had been unpopular with much of the military and ridiculed in the press. A quick and successful demonstration was needed. The Applied Physics Laboratory at Johns Hopkins University adopted a streamlined management approach and within eighteen months delivered a demonstration in the form of the "Delta 180" system for about one-third the cost and time that had been estimated under the conventional system. The Applied Physics Laboratory had shown how the challenge of a hard, short schedule could focus a group to improve its performance and hold down cost. They had minimized the documentation and had emphasized rapid decisions with reduced review. These departures from established form appeared to increase risk, but the program had succeeded. The Space Council coined the slogan "faster, better, cheaper" for this management approach and inserted it into a speech by Vice President Quayle, all in an effort to make the Space Exploration Initiative appear plausible (Foley 1995; McCurdy 2001).

To reshape NASA, the Space Council needed a new administrator as well as a new slogan. Their first choice was the former director of "Star Wars," Lt. Gen. James Abrahamson, but his connections to the military and to the Republican administration developed into liabilities. Eventually, in April 1992, Dan Goldin, a registered Democrat with no direct military connections, was selected. He quickly concluded that the declining funding for NASA dictated a new approach to space science: less management overhead and smaller missions that would utilize more advanced technology. Cost had to be driven down, even if new missions had to assume risks that might have been avoided with the old, conservative management style. He adopted the "faster, better, cheaper" label, even though its prior attachment to manned exploration of Mars might have made it seem a trifle ironic.

Two previous attempts to get NASA to think smaller had been overwhelmed by the internal bureaucracy (Roy 1998). Goldin was going to have to shake the organization to its foundations, and he made the process very public. To some, he seemed more interested in showing that he could carry out a radical reorganization than in defending the agency. On some occasions he even requested less money from Congress than had been set aside in the administration's budget. The Washington system works in an advocacy mode, and an agency that does not state firmly its case for more resources is likely to suffer. Fisk later summed up the problem: "When AXAF was turned down by the Office of Management and Budget, Fletcher [then NASA administrator] took the issue to the president. But Goldin never wanted to throw down the gauntlet." Goldin directed Space Science to operate within a constant budget projection, reversing the optimism and upward pressure asserted by the Fisk plan. That strategy made it impossible for NASA to recover from the budget cuts of previous years. Goldin declared that all future missions would be so small that loss of one would not be sufficient to damage the reputation of the agency. Success would be ensured by developing a large number of inexpensive missions. Failure would be tolerated to keep costs down, and overall success would come through strength of numbers. Goldin also wanted cost growth stopped abruptly and threatened to cancel any mission that grew in price by more than 15 percent.

Goldin put NASA in a state of turmoil that was anathema to Fisk's sense of order. Goldin referred to Cassini, a major mission to Saturn, as "the future of the whole agency mounted on a Roman candle." He made it clear that Cassini was to be the last major mission for the foreseeable future, and his definition of "major" overlapped with Fisk's view of "moderate."

One of Goldin's first acts was to break up Fisk's Office of Space Science. In October 1992, Fisk was appointed to the largely honorific position of NASA chief scientist. Within a few months he left for a faculty position at the University of Michigan. Planetary scientist and cosmic chemist Wes Huntress replaced him as acting associate administrator but was not confirmed in the office until March 1993. To improve its efficiency, Goldin put the agency through a "zero base review." This activity targeted the manned program and the overall infrastructure and generally spared science; still, as the infrastructure was slashed, morale within the agency plummeted.

The combination of Congress's 1991 mission killing and Goldin's reorganization had resulted in a mass extinction: all of the big missions—the dinosaurs—were dying. The hope was that a new type of mission—faster, better, cheaper—would emerge mammal-like from the remains. SIRTF had been one of the weakest of those dinosaurs, curiously ill adapted and continuously threatened with extinction even in normal times. Would it just die? Or could it emerge transformed?

NASA seemed to be spinning out of control. Charlie Pellerin initially flourished under the Goldin regime but soon left NASA for the University of Colorado's business school, where he established a management consulting and training service. His interim replacement, George Newton, told us to think very, very small—to see what we could do with a $100 million mission. Could we somehow split our $1.3 billion mission into two $100 million ones? These questions put SIRTF on the level of some of the smallest NASA science missions, hardly appropriate for one of the Great Observatories.

Eventually Dan Weedman was named permanent head for Astrophysics. Weedman was a strong SIRTF supporter—in fact, to take his new position he had to resign as a co-investigator on the SIRTF Infrared Spectrograph. He was leaving a professorship at Penn State with the specific goal of getting SIRTF started. JPL had named a new SIRTF manager, Jim Evans. Evans was about the same age as Spehalski and also trained as a mechanical engineer, but the similarity stopped there. He seemed to have a permanent smile in place of Spehalski's frown. He listened carefully to the SIRTF scientists, smiling beatifically at their strong views and letting their harangues and squabbles play out until a consensus was reached. His previous experience with science missions was limited, but he had been selected carefully to bring a new atmosphere to the work on SIRTF. However, Evans had undertaken a forbidding assignment: to cut the cost of SIRTF by a factor of 2.5, to less than $1 billion. There was surprisingly little controversy, even though it was clear that we would suffer painful losses to the science we had hoped to achieve.

In addition to the requirement not to irritate Congress with a proposal for another multibillion-dollar mission, we were told to design SIRTF to use a smaller rocket, an Atlas. An Atlas launch would cost about $80 million, nearly six times less than the Titan IV. NASA did not officially count the launch costs in the total for the project; nonetheless, lower launch costs obviously made a mission more affordable. In addition, a smaller rocket meant a lighter satellite, and the first term in cost models is usually the weight. We quickly decided that the 85-centimeter SIRTF telescope was already so small that we could not afford scientifically to let it be reduced. To fit the smaller rocket and budget, we would have to make significant reductions in the SIRTF instruments instead. A subgroup of scientists not involved directly in the instrument construction declared that both the far infrared spectrograph and the longest wavelengths in the imaging photometer had to go. The meeting, held on February 14, became known as the "St. Valentine's Day Massacre." Because Evans had persuaded the scientists themselves to make the recommendations, his ability to lead the project peacefully was not affected. Suddenly, JPL was not only letting the scientists play a central role in the mission, it was doing so in a way that made its management role easier.

After the basic decisions about science capabilities, we needed an engineering breakthrough to advance the new concept. Johnny Kwok, the JPL facility engineer, pointed out that it took several rocket firings to put a satellite into a high orbit around the earth, which meant that some of the potential weight of the satellite had to be replaced with rocket fuel. Using a Styrofoam coffee cup as a prop, he argued that we could launch a more massive observatory if we just gave it escape velocity from the earth, putting it into orbit around the sun. Kwok later had a graduate student work out the details. They showed that the satellite could satisfy all the launch constraints and would quickly get far enough away from the earth to escape its heat, but would take years to get so far away that sending back the data would become a problem. With this orbit, we needed only a minimal load of liquid helium to keep the telescope cold, further reducing the weight. Since we were going to use the much less capable Atlas in place of the gigantic Titan, the twin gains were critical: greater launchable weight for the satellite, less liquid helium.

Even with this breakthrough we struggled to fit everything necessary within the constraints of an Atlas launch. An aesthetically attractive concept resulted. The package had a streamlined look that made the previous Titan-launched version look shockingly obese. The instruments retained a useful and versatile suite of capabilities. And the price was, indeed, just under $1 billion.

The new concept was unveiled in the spring of 1993 in a series of visits to congressional offices. The reception was enthusiastic, a surprising attitude from people who were paid to say no to new science missions. I had an interview in April with newly confirmed Space Science director Huntress, who assured me that the new SIRTF was securely in place in his budget and that he was confident it would be started at the next opportunity. When all of us compared notes after extensive visits in Washington, Huntress's attitude seemed to be nearly universal. We confidently waited for the next federal budget to be finalized so we could finally start building SIRTF.

In August 1993, NASA suffered another embarrassing failure. The billion-dollar Mars Observer needed a small correction in its trajectory toward the planet, but when it was commanded to fire a thruster, it abruptly disappeared. The cause was not obvious. The confusion within NASA was too widespread to keep hidden, and editorial cartoonists and talk show hosts had a new set of jokes at the agency's expense. Eventually, it was decided that a leaky valve had probably allowed oxidizer and fuel to mix in the engine plumbing. When the engine was commanded to start, they exploded and blew the spacecraft apart.

A weather satellite was lost on the same day as Mars Observer. Earlier in the month, a gigantic Titan IV rocket had failed during an air force launch. Space was in the news, and none of the stories were good. Even though the latter two failures were not directly associated with NASA, the distinction was probably lost on the public. Opinion polls had showed a rallying to support NASA after the *Challenger* disaster, but new polls indicated a strong swing in the opposite direction. Nearly 40 percent of the public advocated cutting the NASA budget (Roy 1998).

By the time the budget appeared, the situation in Washington had also changed. The highest priority had become reducing the budget deficit, and various political constraints allowed reductions only in the small portion that contained NASA Space Science among a short list of other programs. The events of August made Space Science even more vulnerable. On September 6, we were told that our lovely version of SIRTF was dead.

5

1993–1994: PICKING OURSELVES UP OFF THE FLOOR

■ We received grim instructions. All new missions were to be cost-capped at $500 million. After the recent painful exercise to get below $1 billion, this new challenge seemed impossible. Nonetheless, I caught the earliest possible flight to Boulder and started working with Ball on yet another IRAS-based bargain-basement SIRTF. Evans was a changed manager, demoralized by the rapid rejection of what had been a brilliant technical and managerial achievement. On October 31, 1993, he was appointed to a new position in JPL and Larry Simmons was appointed to lead us in inventing a new approach to build SIRTF.

At the time, Simmons was leading the completion of the HST WFPC2, a new imager designed to correct the optical problems with the HST. This instrument was installed in late 1993, along with the COSTAR system that dangled correcting lenses in front of three other instruments to bring them close to the originally anticipated performance level. Getting Hubble to work properly was critical to the eventual recovery of NASA Space Science—and with it, our own prospects. Simmons took time from this effort to commission industrial studies of how to build SIRTF for the new low price.

While these studies were proceeding, the scientists also struggled with the problem of how to reduce the mission to about 20 percent of its originally envisioned scale. Clever redesigns would help, but it was clear that we were going to lose more science capabilities. Following a suggestion from Jim Houck, on the weekend of November 6–7, 1993, we held a retreat at Ball Aerospace's Broomfield Conference Center, just north of Denver. The Broomfield Center was like a pleasant if somewhat old-fashioned motel, but with a conference room in the basement. We gathered there after breakfast Saturday morning. The atmosphere was mostly collegial, but with a strong underlying tension. Advocates of each instrument were emotionally armed and ready to fight the reductions. Toward the end of the day, a few of us suggested a different approach: we should decide on a small number of key science programs that SIRTF should be able to complete and then use these to determine which capabilities should be kept. This approach had three benefits: it ensured that we would have a coherent set of capabilities when we were done with the cutting; it gave us decision rules to help avoid loading the mission up with general-purpose capabilities without evaluating their worth, as had happened with science missions before us; but more important for the moment, it allowed all of us to concentrate on something other than what we were losing and to reach consensus on the goals for the mission before we had to trim it down.

On Sunday morning Frank Low showed us a crude sketch of a new design idea for the observatory (Figure 5.1) and suggested that the telescope could be launched "warm"—at room temperature. The idea had come to him late the previous night, and he had finished the sketch at 2:00 A.M. For the previous infrared missions, the telescope had been enclosed in a dewar—a helium thermos flask—and cooled along with the instruments. The telescope and instruments were launched cold, at a temperature just above absolute zero. Because the entire telescope has to be enclosed in the massive vacuum-tight case of the dewar, a cold launch limits the size of the telescope. Low's new approach placed the instruments inside the dewar but left the telescope outside; in other words, only the instruments would be launched cold. The telescope would radiate most of its thermal energy to space during the first month on orbit, after which the helium coolant would take over to pull the temperature down further. Low had realized that the heat capacity of all the telescope materials would be very low once they were cold enough. Thus, only a small amount of liquid helium would be required for the last stage of cooling, so long as other sources of heat were kept away (as was made possible by Kwok's solar orbit).

Low's sketch made the concept look simple and straightforward. Nonetheless, he had carried things too far: to achieve a modest additional advantage in the design, he proposed shrinking the instruments almost to pocket size, far smaller than was compatible with reasonable designs. As a result, his idea was not very popular with this instrument team–centered group, although no one was in a mood to be openly negative.

We spent the rest of the day developing the defining science programs. This list would determine whether each instrument team would prosper or

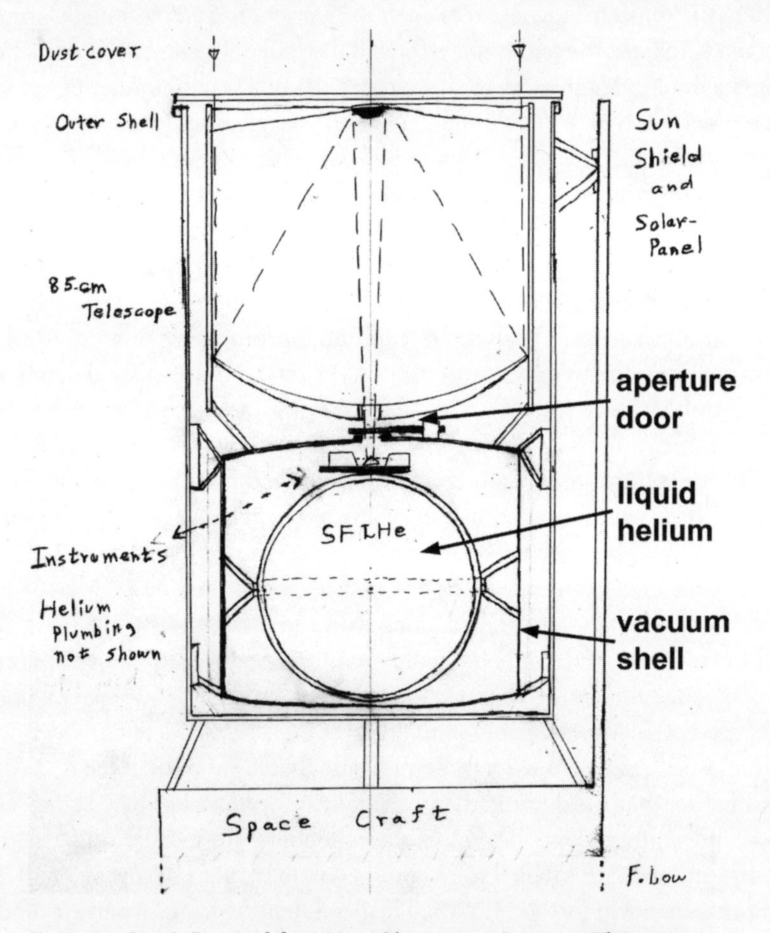

Figure 5.1. Low's Proposal for a New Observatory Concept. The instruments are mounted on the liquid helium vessel inside a dewar (and kept very small to minimize the overall dewar size). The telescope is outside the dewar and cools initially by radiating heat into space; helium cooling is applied only for the last 40 degrees. Some labels have been added. (Courtesy of Frank Low, who re-created the drawing because the original was lost)

wither, and it was approached with suitable intensity. We eventually adopted four areas of research: (1) brown dwarfs, (2) planetary and protoplanetary systems, (3) active galactic nuclei and ultraluminous infrared galaxies, and (4) the formation of galaxies in the early Universe.

Brown dwarfs are objects similar to stars, but they are not powered by thermonuclear burning of gas in their interiors. Thermonuclear reactions, such as the combining of hydrogen nuclei into helium that sustains the sun and gives a fusion "H-bomb" its power, require extremes of temperature and pressure. Objects with less than about 8 percent of the mass of the sun do not develop enough pressure in their cores to sustain these reactions. Thus, they never produce sufficient internal heat and pressure to resist gravity, and they shrink steadily as they lose stored heat from their interiors. Thus brown dwarfs slowly cool and fade as they age, making them hard for astronomers to detect. At the time we adopted them as a science theme, few were known, and even in those cases astronomers debated their identification. Because they are cooler than stars, they should be more easily identified in the infrared than in the optical. We would take advantage of the great sensitivity of SIRTF in the infrared to see if brown dwarfs were an important by-product of star formation.

Brown dwarfs have a family connection to massive planets such as Saturn and Jupiter. These planets also glow faintly in the infrared from energy that was stored in their interiors when they formed. However, the brown dwarfs floating freely in interstellar space are also relatives of the stars because they formed through the same processes that form stars. Star formation begins when a massive, dense cloud of interstellar gas breaks up into clumps due to internal turbulence. These clumps are delicately balanced between their weak (because they are so diffuse) gravitational pull toward collapse and weak (because they are so cold) internal pressure toward expansion. Eventually, something—perhaps a shock wave powered by a nearby supernova explosion—compresses the clumps and gravity takes over, pulling them inexorably inward. From then on, their fate is determined. The force of gravity pulls them into more and more compressed states, which soon reach a stage that can be understood by applying well-understood laws of physics, and which we recognize as various types of star. However, the initial processes in the natal cloud are complex, chaotic, and not well predicted by physical calculation, much like the complexities in our atmosphere that result in weather. The course of these initial processes is imprinted in the final range of masses of the stars and brown dwarfs that emerge. In the days when we adopted the brown dwarf theme, theories for the turbulence in clouds and the eventual collapse into stars predicted that there would be a lower limit to a collapsing clump, close to the lowest-mass stars observed. Finding where this limit lay would be a basic test of these theories and would give us clues to how stars in general formed.

The *planetary and protoplanetary systems* theme would study dusty disks of planetary debris surrounding some stars, which warm them and make them glow in the far infrared. We think these particles may be the result of devastating collisions between small planets that shattered them back into tiny particles. These "debris disks" are by far the most dramatic signposts of other systems of planets outside the solar system. The reason has to do with surface area. The earth, for example, has only one-millionth the surface area of the sun. Even if the earth and sun were at the same temperature, the earth would thus radiate only one-millionth as much energy as the sun. A similar planet around a nearby star would be virtually undetectable in the glare of the more-than-a-million-times-brighter star. However, suppose the earth were broken up into a trillion trillion particles. If they were all the same size, they would each be about one-fourth inch in diameter, and their total surface area all added together would exceed that of the sun by nearly a million times. If the star they orbit heats them up to earthlike temperatures, they suddenly become quite easily detectable in the infrared. In fact, the first example was discovered by accident with IRAS when it observed what was thought to be an exceptionally well-understood star for the purpose of calibration. Our plan was to probe many more stars than could be reached by IRAS. We would use the greater sensitivity levels and more sophisticated instruments on SIRTF to understand exactly what debris disks were telling us about the number of planetary systems in the space around us and how these systems had evolved from the time of their formation.

The *active galactic nuclei and ultraluminous infrared galaxies* program was to determine the connection between these two types of objects. Active galactic nuclei (AGNs) have mystified astronomers since their discovery in 1943. We now agree that they result when interstellar gas falls into a supermassive black hole (a billion times the mass of the sun). When material disappears into a black hole, much of its mass is converted into energy through Einstein's famous relation $E = mc^2$ (energy equals mass times the speed of light squared). Dropping it into a black hole is a far more efficient way to convert mass to energy than the hydrogen fusion that powers normal stars.

Most of the ultraluminous galaxies are strongly interacting systems comprising two large spiral galaxies that have collided and are merging into a single system. Computer models show that the interstellar gas and dust can shed energy rapidly in such collisions, and hence tend to spiral inward toward the center of mass. The resulting concentration of dust hides the activities in the new galaxy's nucleus from optical astronomers, so the standard tools for assessing what is happening do not work well. The large infrared outputs make it clear that a lot of energy is generated, but the dust absorbs nearly all of it and releases it as heat radiation in the infrared.

By the time we adopted this theme, it was clear that these systems were powered by some combination of very recent formation of very massive stars, probably triggered by collisions and shocks in the gas clouds as they settled into the centers of the merging galaxies, and AGNs, probably as a result of some of that gas settling all the way into the core where a previously existing black hole was waiting to suck it into its immense gravitational field. However, debates raged about the relative importance of the two processes. There was widespread speculation that in the ultraluminous galaxies we were seeing a part of the sequence of events that leads to formation of an active nucleus.

The fourth theme was to probe the *formation of galaxies in the early Universe*. As a result of the expansion of the Universe, we see ever more distant objects moving away from us at ever increasing speed. As the Universe expands, it stretches the light emitted by distant galaxies, increasing its wavelength and thus moving it redward in the electromagnetic spectrum. For example, this cosmic redshift moves the optical region further and further into the infrared. We describe the amount of shift as z, the ratio of the change in wavelength to the original wavelength. Consider a distant galaxy in the deepest images by the Hubble Space Telescope, such as the Hubble Deep Field, at a redshift of $z = 3$. When we look at this galaxy, the light we now see in the red at a wavelength of 0.6 microns was originally emitted at $0.6 \div (1 + z) = 0.15$ microns—well into the ultraviolet. Because it takes so long for their light to reach us, objects we see at high redshift are young; those at $z = 3$ are about 25 percent of the way from the Big Bang to the present. Astronomers are fascinated with probing the properties of galaxies at this early stage of development.

In general, stars form in dense clouds full of dust that is extremely efficient at absorbing ultraviolet and visible light from the new stars, whereas it has little effect on infrared light. To get a better look at the stars in dusty galaxies at redshifts near $z = 3$, we need to go to their near infrared emission, 0.8–2 microns, which is now received by us in the 3–8-micron range. In addition, the absorbed ultraviolet energy will heat the dust and be emitted by it in the far infrared. Thus, a series of deep surveys is required to fill out the picture of early star formation and assembly of galaxies, similar in philosophy to the Hubble Deep Field but concentrating on the 3–8-micron and far infrared ranges. These observations together would complete the picture of galaxies in the early Universe we are getting from HST and from giant ground-based telescopes.

We adopted these four themes and briefly considered the minimal instrument capabilities needed to pursue them. Virtually all moving parts would have to be stripped from the instruments (an exception was eventually made for a shutter in IRAC [Infrared Array Camera] and a moving scan mirror in MIPS [Multiband Imaging Photometer]). Instrument versatility usually depends on mechanisms that can bring members of a suite of spectral filters into the beam or scan the wavelengths in a spectrometer. We would have to depend on the power of our infrared arrays (see Figure 5.2) to compensate for the loss of versatility. Having committed to this path, we scrambled to the Denver airport to get home the evening of November 7.

We got together again at JPL on December 2 and 3. It was a rough meeting. Low demanded that the project switch immediately to his new concept. With his industrial studies still in a very preliminary state, Simmons was cautious. Low was adamant and argumentative. Simmons became more and more defensive. Yet, some decision had to be reached. At the end we all agreed to pursue the new idea as vigorously as possible to see if it really would work. Since the cold launch was a known approach, we felt we could set it aside for the time and concentrate on Low's idea.

The industrial work was reported at a Science Working Group meeting in mid-December. The studies had been done very quickly and there had been little time for creative thought. The cost estimates were in agreement with the new guidelines, but we did not take them at face value. The designs represented a retreading of old concepts, and none of us were convinced that they offered any real potential for a lower-cost mission. The meeting was a depressing reminder of our difficulties, but it also increased our commitment to pursue Low's new concept. Low had produced a new sketch that turned out to be remarkably similar to the final version of SIRTF. In a rare departure from his sense of design aesthetics, he had even slightly increased the space allowed for instruments.

More than anything else, a struggling NASA project needs a picture of its mission concept. Without a picture, it just isn't possible to convey any sense of reality. Even though we had not made some basic decisions, Simmons decided that we needed to assemble a review panel as soon as possible to authenticate the new approach. If the review was positive, we would put together an engineering description that would provide the picture we needed.

We were ill-prepared for the panel that met on February 14, 1994, but the panel members had been intentionally selected to be friendly. Nonetheless, their assessments were mixed. Jerry Smith, who had played a central role in managing IRAS, summarized: "I believe it can be done, but only because of our experience with IRAS. The presentations today did not convince me." Other panel members felt that we lacked a coherent concept and that we needed to consider radical departures from traditional NASA management approaches to live within the new budget.

In addition, there were concerns about whether the new version of SIRTF,

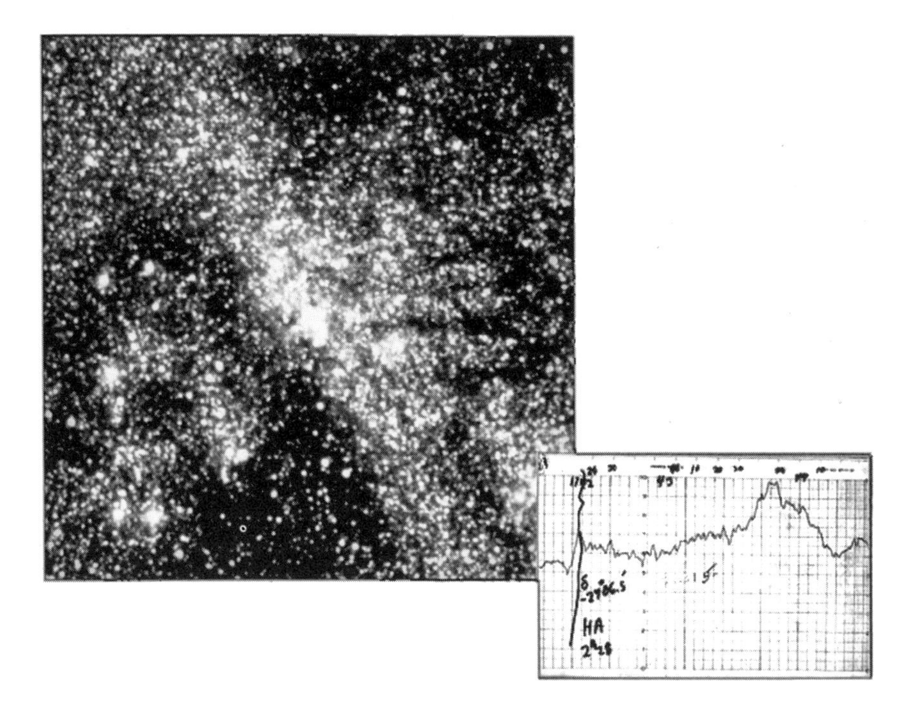

Figure 5.2. Power of Infrared Arrays. At lower right is a strip-chart recording of a single infrared detector scan over the nucleus of the Milky Way. Crude maps were painfully constructed by many such scans; the observing time for these maps is the exposure time required times the number of resolution elements in the map and often extended to many nights of observing. With an infrared array, as shown in the upper left, a single exposure with as many as a million detectors can give a detailed image of the same region virtually instantaneously.

with its reduced capabilities, could still be identified with the mission that had been endorsed by the Bahcall committee. A National Academy committee was to review our merit as the top recommendation for the decade. We were to present our case to the committee on February 17; they would also hear a presentation for the proposed airborne infrared observatory, SOFIA (the Stratospheric Observatory for Infrared Astronomy).

The scientists involved with SIRTF and SOFIA would normally have felt some allegiance. We had worked together on similar science questions, and many of the SIRTF scientists had used SOFIA's predecessor, the Kuiper Airborne Observatory. The slim prospects for both projects made everyone defensive, however, and our group even more so because the meeting was held in a conference room at Ames Research Center, the home center for SOFIA. The room was plastered with posters advertising SOFIA, and any empty chairs were soon occupied by people working on that project, despite the ruling prior to the meeting that each project could have at most three representatives. Our paranoia increased when Jim Houck was disqualified from the panel for conflict of interest, yet prominent users of the Kuiper Airborne Observatory and known partisans for SOFIA were left sitting. Mike Werner, Larry Simmons, and I tried to set these worries aside while we presented our case. We then waited apprehensively. After a number of weeks, the Bahcall recommendation was reaffirmed. SIRTF was described as the "highest" priority and SOFIA as a "high" priority.

Over the next few months the project engineers continued to study Low's warm launch idea, or rather their version of the idea. He met with them frequently, but many of the meetings ended in noisy arguments about areas where they had departed from his concept. Low has a brilliant "feel" for technical solutions that allows him to reach many correct insights (and a few incorrect ones) intuitively. As a result, the intermediate steps may be missing in his thought patterns, making discussion of the details difficult. Eventually, however, the engineers announced that they could find no serious faults.

There was never a rigorous engineering tradeoff between the new concept and the older "cold launch" one. Cold launches had proven successful for IRAS and a number of smaller infrared telescopes, and were about to be demonstrated again with ISO. At the SIRTF aperture of 85 centimeters, the cold launch was still the conservative route from an engineering standpoint.

Low's idea also indicated a direction for larger future infrared telescopes, for which an adequately sized dewar would be impractical (NASA's plans for the 6.5-meter-aperture James Webb Space Telescope [JWST] revolve around a warm launch telescope). It made full use of Kwok's solar orbit to get extraneous heat away from the telescope. Together, the two ideas had the immense political advantage of a fresh start for SIRTF that showed we were not just nursing the same mission we had begun with years before. Very few of the engineering details can be expected to catch the fancy of those making life-and-death decisions about a mission; broad impressions like creative new mission concepts have a much better chance. John Bahcall recognized that when he noted that "SIRTF survived because of the innovations introduced by the scientists involved with the mission." Goldin was pushing for new approaches to enable outstanding science within reasonable budgets. Years later he recalled that "SIRTF was saved by a couple of kids at JPL who came up with the idea of a solar orbit." Although the description of Kwok and Low seems wildly inaccurate, he says he meant it as a compliment to their mental agility.

The possibility of a cold launch SIRTF was soon eliminated altogether by a fiat from NASA Headquarters that we would use a small Delta launch vehicle. Weight was now our biggest concern, and the warm launch became a necessity. A team of JPL engineers set out to assemble a full description of a new, very slim version of SIRTF. We carved and squeezed to fit everything into it. The recent turmoil meant that no one had a very advanced design, so we had to use guesswork to find an allocation of weight among the systems that we thought would be feasible. The resulting study was called the "Greenbook" after the color of its cover. With the technical backing of the Greenbook, we produced a number of pictures, such as the one in Figure 5.3. The illustration was based loosely on a previous attempt in which we had shown that the Titan-SIRTF related to HST about as the Atlas-SIRTF relates to the Titan one in this picture. However, the self-referential version carried a more powerful message: we had brought down the size, mass, helium volume, and cost of the mission substantially, with only a modest compromise in mission lifetime.

But had we gone too far? Could we actually live up to what the picture promised? Sudden relief from our worries came from an unexpected direction. The Japanese space science agency had decided to launch an infrared telescope shortly before SIRTF and wanted to coordinate that mission with ours. NASA elected to explore shifting one of the SIRTF instruments to the Japanese telescope. Doing so would remove enough weight to make SIRTF fit within the Delta capabilities, and also would provide the type of insurance Goldin was seeking for high-visibility missions. By splitting the mission, NASA could proclaim a success if either of the two succeeded. No more Hubble/Galileo/Mars Observer disasters, which could wipe out the goals for an entire area of science, were to be permitted.

The new opportunity came with its own risks and tradeoffs. The Japanese space program and NASA operate in very different ways. The Japanese mission had a strict cost cap equivalent to $100 million, and its capabilities would be cut ruthlessly to stay under this limit. It would use a new rocket, the M5. Given the projected weight of their telescope, this rocket could manage only to put it into a fairly low orbit around the earth. To compensate for the resulting warm environment, the Japanese were planning to use mechanical coolers and to restrict severely the area of the sky that the telescope could view at any one time.

But still, we had waited so long that any possible ride into space seemed worth serious consideration. The two candidate instruments were Giovanni Fazio's and mine. Each of us spent long hours looking at the tradeoffs and talking to the Japanese scientists. At one point, everyone involved gathered in Paris for a conference on future infrared space astronomy missions. I engaged in long discussions with Tetsuo Matsumoto, a senior Japanese astronomer, over dinner at Le Moulin Verd, a very traditional French restaurant around

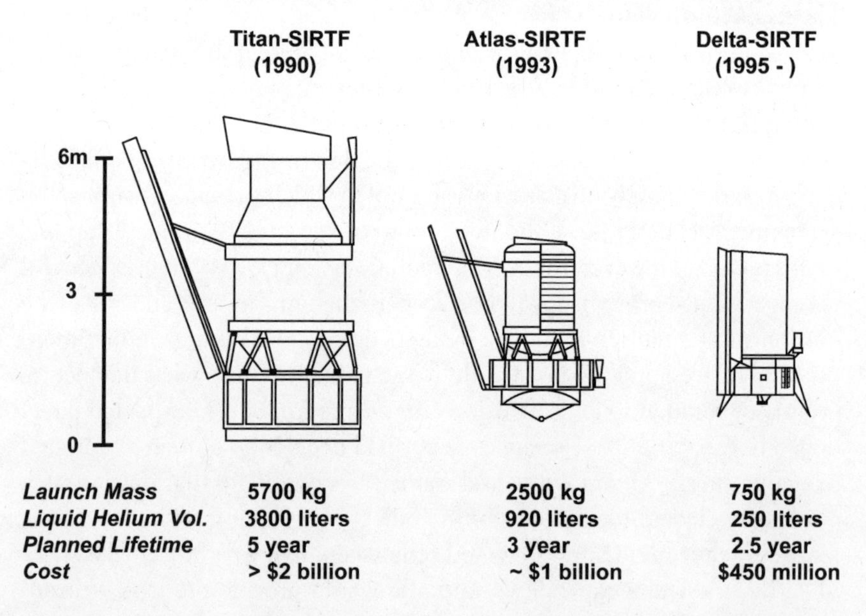

Figure 5.3. Evolution of SIRTF. This figure was one of the critical elements in finally getting the mission started.

the corner from the budget hotel where all the conference participants were staying. Between courses we sketched mission concepts on the paper tablecloth. I was torn. A part of me wanted to switch to the Japanese mission, but the requirements for very low temperatures and the other constraints associated with my far infrared instrument made the risk immense. The long, frustrating wait sometimes overrode cold reason, making the gamble tempting even if the stakes were high and the chances of winning low.

After a few months it was decided that IRAC would become Japanese, and that management control would be shifted to Goddard Space Flight Center. Management by a NASA center was considered to be necessary to protect the agency's interests in a foreign collaboration. This decision downgraded the role of Fazio as principal investigator. He was very upset at the change in his status, but soon he and the rest of the IRAC team were working hard with Goddard to make the Japanese-American mission a reality.

Meanwhile, the prospects for new missions went from bad to abysmal. The federal budget had been running up huge deficits since the early 1980s, and Congress had imposed draconian spending caps on itself to force greater discipline. Projections of the Space Science budget for the next five years showed a steady decline to a total loss of 15–20 percent, more if any allowance was made for the decreased purchasing power of future, inflated dollars. The problems with three major missions in rapid succession—Hubble, Galileo, and Mars Observer—had cast a pall over the entire program. The Office of Management and Budget (OMB) was thinking seriously of a permanently reduced national investment in space science. Huntress was struggling to find a way to keep the operating missions going and to finish the ones already under development, and even that limited goal showed him running a deficit soon. Goldin was still busy shaking the management of NASA to its foundations. There seemed to be no limit. He had even pushed to the front of the queue a new mission to Pluto sold to him by an enterprising engineer who had brought his viewgraphs to a cocktail party specifically to gain access to Goldin. There was a lot of black humor about walking by NASA Headquarters very carefully so as not to get hit by managers who had jumped off the roof.

We continued doggedly to push ahead with the now more easily launched two-instrument SIRTF. We struggled to make our plans fit into a $400 million budget while IRAC and Goddard made rapid progress. All of us met periodically to coordinate, since we were still considered by NASA to be a single mission divided only for programmatic reasons—a single mission with two launches on opposite sides of the earth.

At the February 1995 Science Working Group meeting there was a series of presentations by Goddard representatives on the Japanese initiative (there was no one from Japan at the meeting). Goddard had taken over Low's warm launch concept and was promising to expand the capabilities of IRAC far beyond those that would have been possible on SIRTF, in particular by providing it with very large detector arrays that were just entering development and were not yet available. (As it turned out, only a few of these arrays had been produced by the time IRAC was completed five and a half years later, and an IRAC using them would have been seriously delayed.) The estimated American share of the cost for the Japanese-American mission was well above the amount allocated by NASA Headquarters. An IRAC presenter had accidentally included a viewgraph that laid out how the Japanese-Goddard-Smithsonian combine would build a mission that would carry out all four of the SIRTF defining science goals, with no need for SIRTF itself.

This radical makeover of the Japanese mission had been carried out without the close involvement of the Japanese, who understandably had no interest in seeing their mission taken over by Goddard. In March, we were told to consider the possibility of IRAC rejoining SIRTF. Over the next couple of months we struggled with reintegrating IRAC onto an already overloaded rocket, along with the other implications of restoring SIRTF to its previous form. The final decision that IRAC would rejoin SIRTF was announced at the end of June. Fortunately, a more powerful form of Delta rocket became available soon afterward, so we no longer had a problem with the extra weight; the other observatory resources, however, were going to be stretched even thinner than before. The extra $50 million we were allocated for IRAC and the associated costs proved to be insufficient. Furthermore, the change in the management relationship between Goddard and Fazio as instrument principal investigator was not reversed, even though the ostensible cause for Goddard to be in charge had been removed.

Perhaps it wouldn't matter. It looked like IRAC had been loaded onto an already sinking ship. A preliminary NASA FY 1996 budget had shown Space Science being cut by 18–20 percent and Astrophysics by 12–15 percent, but it had still been possible to squeeze in both SOFIA and SIRTF. Then there had been an election. Congress was taken over by Republicans anxious to show their fiscal conservatism, and they imposed further cuts to allow a tax reduction. NASA was trying to handle the $1 billion per year impact by reducing infrastructure—staffing at NASA Headquarters was being cut in half—but that only symbolized the scale of the problem. In addition, in the budgetary confrontation of congressional Republicans with the Democratic executive, the federal government was shut down for a number of months, making the budgetary process emerge in slow motion long after the budget was supposed to be approved.

6

ISO SHOWS THE WAY

■ In November 1995, ISO—the first cold, pointed infrared observatory—was launched by the European Space Agency. It operated nearly flawlessly for two and a half years and made many important discoveries. To remain viable, SIRTF was going to have to provide capabilities well beyond those of ISO, even in the face of our stringent budget limits.

The ISO detectors were generally a couple of generations behind those to be used in SIRTF. There were two true detector arrays, both 32 × 32 pixels in format. The remaining channels were all based on small numbers of detectors connected individually to readout amplifiers. Many of these devices gave excellent performance on a per-detector basis, but the observing efficiency was low due to the small number of detectors. On the other hand, the ISO instruments had a broad variety of features implemented by complex mechanisms and optics of the sort we had been required to cut out of SIRTF. These features worked well and enabled a versatile suite of science capabilities. Much of the ISO science came through spectroscopy, which included a broad range of capabilities that extended well beyond those we had been able to retain in SIRTF. For example, ISO opened new vistas on the molecular content of the interstellar medium, finding water vapor and molecular hydrogen in a broad variety of environments. ISO contributed greatly as well to the four defining programs we had identified for SIRTF.

Sometimes science moves forward by leaps and bounds, and such was the case for brown dwarfs by the time of the ISO mission. ISO maps of star-forming regions found many low-luminosity, young brown dwarfs, but much of these data had been anticipated by ground-based telescopes. Detailed studies of star-forming clusters in the near infrared (near 1 micron wavelength) had succeeded in determining the temperatures of the objects in a different way. Certain atoms and molecules imprint a signature of the stellar temperature at the specific wavelengths where they can absorb and emit energy. Measuring these telltale features with near infrared spectrographs had located large numbers of relatively low temperature objects. Combining temperatures with measurements of their brightness made it possible to calculate energy outputs and identify which ones were true brown dwarfs. They were being found in about the numbers expected from extrapolating the counts of stellar-mass young stars downward. In addition, a number of surveys in the near infrared had found many older brown dwarfs among the stars in the neighborhood of the sun, again in the numbers expected if there was no discontinuity in the rate of formation versus mass. Thus, all the theoretical predictions that brown dwarfs would *not* form in significant numbers had to be abandoned.

Sometimes science progresses at a glacial pace compared with expectations. A shortfall in the performance of ISO in the far infrared resulted in the discovery of relatively few new examples of stars with planetary debris systems. Much of the information about the incidence of debris disks and their general behavior still depended on IRAS data. Teams trying to draw even the simplest conclusions from the merged IRAS and ISO data ended up with contradictory results. For example, one team studying eighty-four stars roughly similar to the sun reported that "most stars arrive on the main sequence [where they are sustained by hydrogen fusion] surrounded by a disk; this disk then decays in about 400 Million years" (Habing et al. 2001). Another team (including some of the same people!) reported in contradiction that "the 5 stars in our sample [of 30] which have an infrared excess are probably older than 3 Billion years" (Decin et al. 2000).

ISO did advance this research area in another direction. Spectra of very dense and prominent young disks established the presence of crystalline material in their silicate-rich particles (silicates, compounds based on combinations of silicon and oxygen, are among the main constituents of the earth, including common rocks). The silicate materials in interstellar dust are in an amorphous form that can be converted to crystalline form only by heating. Thus, it seems logical to conclude that the particles in the circumstellar disks must have gone through an episode of substantial heating, and one plausible way for that to have occurred is during the building of planets.

ISO spectroscopy also advanced our understanding of active galactic nuclei and ultraluminous infrared galaxies. When emission lines were measured in a number of ultraluminous galaxies, a surprisingly *small* number of the galaxies showed the very highly excited atomic states that would indicate a hidden AGN in their center. The huge energy outputs from most of them appear to be generated by young, massive stars. The sensitivity limitations of the ISO spectrograph limited the emission-line observations to a handful of objects, however, preventing absolute certainty.

IRAS had discovered that the sky is lit up by warmly glowing dust, dubbed "infrared cirrus" because of the way it looks in sky images. Another mission, the Cosmic Background Explorer (COBE), had observed the sky to wavelengths well beyond the 100-micron limit of IRAS. At about the time of the ISO launch, painstaking reductions of the COBE data had finally succeeded in removing the foreground clutter of infrared cirrus and it had become clear that COBE had discovered an additional far infrared component in the sky that is the sum of the outputs of all the infrared galaxies. The energy in this component is as large as, perhaps even larger than, the total energy emitted directly by all the stars in distant galaxies and appearing in the ultraviolet, visible, and near infrared. That is, very deep images of the sky such as the Hubble Deep Field would never be able to find more than half of the total activity in young galaxies. To find the rest requires deep surveys in the middle and far infrared with telescopes with enough resolution to isolate the individual components of what appeared with COBE's blurry vision to be a diffuse glow. Large amounts of observing time on ISO were devoted to surveying "blank" regions of the sky to detect this population of infrared-bright galaxies.

The longest band accessible by ISO's imaging array was at 15 microns. ISO was able to image galaxies efficiently in this band up to a redshift of $z = 1$. At $z = 1$, the Universe was about 8 billion years younger than now, about 5 billion years post–Big Bang. We can expect the much younger galaxies we see at this redshift to differ from those around us. For even younger galaxies at higher redshifts, $z > 1$, the peak of the galaxy emission curve moves into the longest ISO band at 175 microns. Large amounts of time were spent to make deep surveys there also.

Combined, these two sets of data revealed that the rate of star formation near z = 1 appears to be higher than had been deduced by optical means. At larger redshifts, the behavior of galaxies had to be fitted empirically, using simple trends in the rates of appearance of different types of galaxy to explain the optical and infrared data consistently. A general characteristic of these fits was a rapid increase in the number of luminous infrared galaxies as one looked back beyond z = 1 toward a younger Universe. This result was in good agreement with expectations that galaxies would be colliding and merging frequently, a process that produces many of these sources. However, because the fits generated by different astronomers disagreed substantially on the exact behavior, it was clear that they were not unique. They needed to be tested by filling in the spectral coverage between 15 and 175 microns.

While ISO embarked on its voyage of discovery, we struggled to get started on SIRTF. In December 1995 I visited the Office of Aeronautics, Science, and Technology (OAST) with other members of the Space Science Working Group—a loosely knit organization representing universities with interests in NASA science. OAST oversees the health of the nation's civilian science and technology and hence tends to be an advocate for science. The discussions were nonetheless grueling. A steady decline in Space Science's purchasing power was projected, amounting to some 30 percent over the next five years. Given the existing obligations, including continued operations of space missions already started, such a budget would result in nothing new in the NASA program at the end of this period. The group at OAST listened sympathetically to our arguments and gave us some advice on carrying them further, but there was little they could do. Our fate was in the hands of a Congress led by budget-slashing conservatives.

To counter the grim prospects, Space Science head Wes Huntress had been saving money by patiently and steadily pressuring the ongoing missions to be more efficient. Large savings in the HST budget were implemented shortly after the servicing mission that restored the basic performance of its optics. To take advantage of these savings, however, Huntress had to get approval for new missions to put in the budget. The sentiments at OMB to reduce Space Science permanently after losing Mars Observer had been softened by Goldin's emphasis on smaller missions. Huntress proposed a package of modest missions that were compatible with the new direction.

We had some difficulty getting into this package because different elements of the infrared community were backing SIRTF and SOFIA. Selecting one over the other might polarize this already small community and undermine any chance of either mission moving ahead. This issue was solved when the airborne astronomy community agreed to free up resources for SOFIA by decommissioning the Kuiper Airborne Observatory. Our picture (Figure 5.3) showing the progressive shrinkage of SIRTF also helped, because it demonstrated how far the mission had been descoped. Both actions were in the direction Goldin was pushing the agency—to turn old things off to make room for new ones, and to reinvent large missions as small ones.

New missions are evaluated and approved during a series of negotiations that start in a NASA Headquarters retreat. Huntress describes the atmosphere as "a smoke-filled room full of aggressive, wolflike associate administrators fighting over money. It is very debilitating. Once you've been through it you want to forget it, but of course you have to do it the next year again." He claims that the retreats were so unpleasant that he has selective amnesia and cannot remember any of the details. However, in August 1994 he had been able to get internal NASA approval to put SIRTF and SOFIA up together for New Starts.

The next step after the retreat has set the internal NASA priorities is to work with OMB in September to get the projects implemented in the administration's budget proposal. The OMB budget proposal is passed back to NASA in November, and there is an opportunity for negotiation at that time. Huntress had been aided in getting his missions through this step by Steve Isakowitz, then at OMB. Isakowitz had once been an engineer at Martin Marietta, where he wrote the definitive handbook on space launch systems. Thus, he had a strong personal interest in space exploration.

After OMB, the proposal goes to Congress for approval and funding. With the Republicans now in control, Marcia's strategy of talking to them all along suddenly looked brilliant. Although we occasionally had to bite our tongues hard to avoid saying something unflattering about the "Contract with America," we knew whom to visit, and they knew who we were when we did. Amazingly, Isakowitz went to Congress too, in a previously arranged sabbatical year away from OMB designed to broaden his experience in the budget process by working for Stephan Kohashi as the lead staff member for science in the Senate. It was an improbable coincidence that a highly sympathetic person would receive the NASA Space Science program at OMB just when Huntress had put it in order and won approval at the NASA program retreat, and that the same person would move to Congress to help see the program through there. Perhaps it was the law of averages righting itself after all of our earlier misadventures in Washington.

Although we had finally had some unexpected luck and were doing our best to promote the program, the atmosphere for congressional approval was grimly negative. The SIRTF money was deleted in the House. Isakowitz and Kohashi got it restored in the Senate. Eventually, their position prevailed in conference. Despite the chaos of the government closings and the grim forecasts for the future of space science, we had $10 million to get started!

The other programs in Huntress's package, including SOFIA, had also been approved. Huntress summarized the situation in a "Dear Colleague" letter distributed on March 29, 1996:

> For all of NASA, 1995 was a tumultuous and difficult year. It was also a pivotal year for Space Science in terms of our plan to recover from earlier setbacks and to establish a firm basis on which to build for the remainder of the decade. This required starting new programs in the FY 1996 budget in the face of a newly elected Congress determined to reduce Federal spending. While this Congress was initially resistant to introducing new programs, we worked with them successfully to fulfill our FY 1996 budget goals. This was achieved by presenting them with a Space Science program plan responsive to the country's need to reduce the budget that still preserves an exciting future in a declining budget situation. This could not have been done without unified support and action by the Space Science community.
>
> All of this was accomplished in an unsettling climate of budget and workforce downsizing, coupled with a revolution in Federal business processes. . . . NASA Headquarters faced a 50 percent downsizing requirement over 4 years with an attendant reorganization and new business plan for the Office of Space Science. Finally, a significant portion of the Federal Government was furloughed twice over the holiday season.
>
> It is not possible for me to convey to you the impact all of this had on NASA employees and the heroism that was required of the people who work so hard for the benefit of the Space Science enterprise. We are now back in business and officially reorganized to meet the new staffing targets by 1999. For me it seems most remarkable that in spite

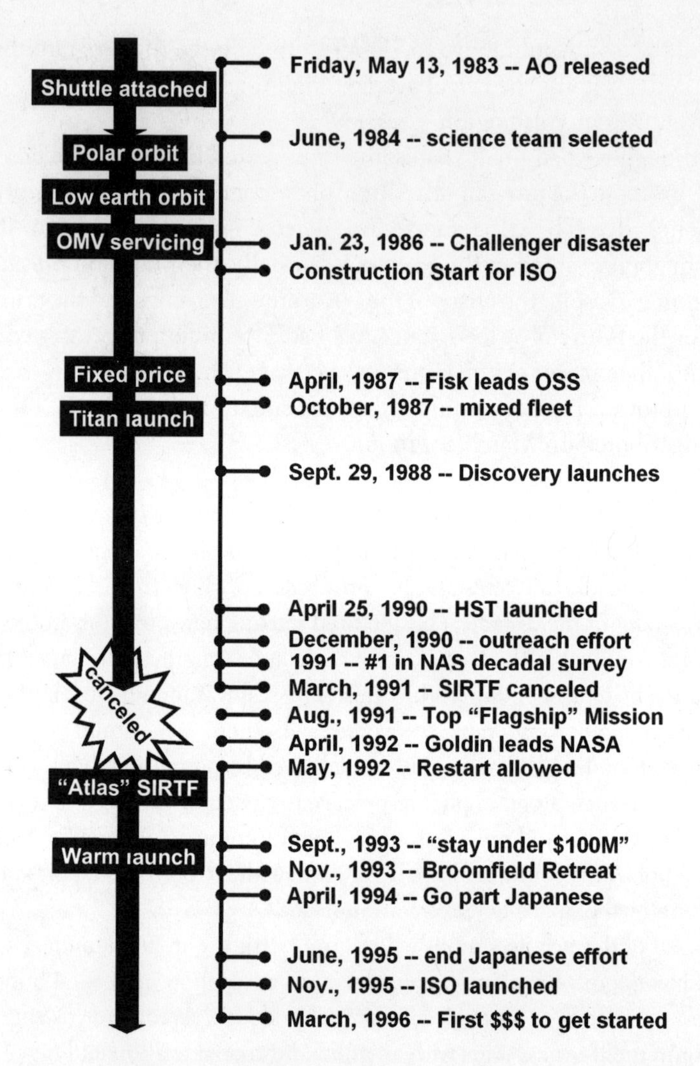

Figure 6.1. Our Road to Getting Started. Major events are indicated above the time line, and different versions of the SIRTF mission are to the left of it.

> of the turmoil, Space Science has been markedly productive in 1995. We should all take pride in the parade of discoveries and new knowledge that has come from Space Science programs. . . .
>
> Other big news in the FY 1996 budget besides continuation of current programs is the approval of starts for SOFIA and the New Millennium program and approval of definition studies for SIRTF and Rosetta/Champollion. . . .
>
> With this message, I hope to convey to you that in spite of a great deal of change and uncertainty, this Nation's Space Science program is performing very well. We have managed to build the foundation for a robust and exciting future throughout the remainder of this decade by responding flexibly and innovatively to the challenges and changes that have confronted us.
>
> . . . The budget just released by the Administration for next year is sufficient to continue all of our current programs through FY 1997, including those started in FY 1996; but we have not yet secured approval for the remaining elements of our Strategic Plan for this decade. The outyear budget future for the Agency and for Office of Space Science remains uncertain and represents a challenge for us over the next several years.

Figure 6.1 summarizes the long road that finally led to getting the funds to start SIRTF.

7

GETTING UNDER WAY

■ In May 1996, twelve years after submitting the initial SIRTF proposals, we met at NASA Headquarters with real money to spend. Congress had committed $10 million to SIRTF in the budget for that fiscal year (October 1995–September 1996), and NASA had matched it with another $5 million. Delivery of these funds had been delayed by the multiple closings of the federal government the previous winter. Finally, however, we were preparing to move SIRTF from Phase A status, when general mission concepts are evaluated, into Phase B, when a specific design concept is to be completed, priced, and presented for approval for construction.

Jim Houck had been working with Ball Aerospace to demonstrate that his IRS could be manufactured cheaply. The Ball fabricators could maintain tolerances so well that the instrument could be assembled in alignment without a sequence of measurements and adjustments. He had termed this approach "bolt and go." Ball was building a demonstration spectrograph that Houck was planning to use on the Palomar 200-inch telescope, and Houck had brought its housing to the meeting to show us. Toward the end of the meeting, Houck and NASA Astrophysics head Ed Weiler took it upstairs to show Goldin, to demonstrate a faster, better, cheaper approach to building SIRTF.

"He said, 'It's a national tragedy SIRTF will only last two and a half years,'"

Houck reported when they returned. "'You should find a way to make it last five years, maybe seven. See what you can do with a bigger rocket.'" I looked around the room at the rest of the Science Working Group. Their faces looked puzzled, even annoyed. "What do we think?" asked Mike Werner. Each of us in turn grasped the proposition cautiously. As I listened, a sense of absurdity slowly asserted itself. For the first time in more than a decade we were being asked to try to improve the mission, even if it cost (a little) more. We were reacting with well-practiced paranoia. Eventually, we cautiously responded to Goldin's pressure by promising ourselves to refine the design of the mission to extend its life. However, any improvements would be hidden from sight. There would be no outwardly apparent changes.

Before Goldin's instructions had interrupted us, our meeting had been focused on another unfamiliar problem. We were scheduled for an external review in a month. After twelve years we were still unpracticed in them, a reflection of SIRTF's peculiar difficulties in converting its promise into reality. The Preliminary Non-advocate Review (PNAR) would judge whether we had laid an adequate foundation to spend the Phase B money wisely. According to the NASA handbook on project development, we should demonstrate that our mission addressed a high-priority science goal for NASA, that we had found a feasible mission concept that could reach this goal, and that our plans were realistic in terms of the scope of the mission and its potential cost. The assignment did not sound difficult. Our meeting was engaged in organizing and fine-tuning material we had repeatedly presented to ourselves in various forms.

But we were still a bit apprehensive. "Non-advocate" reviews assemble expert panels who have no association with the project, and hence should have no reason to advocate its approval. A panel member might seize on some weakness that familiarity had made invisible to us—that, after all, is the purpose of such a review. Or, since they had little familiarity with the project they were reviewing, panel members might develop some mistaken idea. It would have the same impact as a real error unless we could set them straight promptly. Furthermore, at its core, what we were proposing was unconventional and might therefore appear to be flawed to the panel members, whose experience was generally derived from the previous generation of NASA projects.

Traditionally, large NASA projects selected two industrial contractors early in Phase B to carry out parallel studies of the mission. At the end of Phase B, the contractors would submit new proposals. Only one of them would be selected for Phases C and D, detailed design and construction. The competition in Phase B would yield a thoroughly specified design. The contractor would then build to these specifications in Phases C and D. A complex set of rules linked the contractor's profits to his performance in meeting the specifications. Although the emphasis on price competition and rigid development rules was originally expected to be helpful in containing costs, the older missions typically ran to twice their expected budgets. Missions were now required to stay within budget even if they had to sacrifice performance. Severe penalties were threatened for not doing so: Goldin had announced that any mission more than 15 percent above its budget would be considered for cancellation.

Building our mission within the new constraints required us to embark on a management experiment. Simmons would have to cut costs severely and also show that the new bargain basement price was credible. To minimize costs, he had decided to initiate the competition among the potential contractors *before* we entered Phase B. We would finalize the mission concept cooperatively with the selected contractors and carry out Phases B, C, and D in a team relationship with no subsequent competition. Although the cost estimates would not have the benefit of the additional understanding traditionally gained during Phase B, their mere existence would make the total mission cost seem credible. Still, such an experiment was exactly the kind of overriding issue that could make a review board exercise its veto.

Management experiments had become the rule at NASA, necessitated by declining budgets. *Science* magazine described the situation in a 1995 item:

> New managers, many from academia, are setting out to reform and improve [NASA's] research complex, which spans a dozen centers and includes more than 2000 employee and contractor scientists. The changes are a response to a looming $2 billion cut in NASA's $14 billion budget over the next 5 years. . . . NASA Administrator Daniel Goldin wants to reduce personnel rather than cut programs to accommodate the budget squeeze. Science managers, with their nearly $3 billion research portfolio, are eager to be seen doing their part. But they face members of half a dozen disciplines within NASA and in the academic community who hunger for new projects that are incompatible with a shrunken agency. ("Space Science: Will NASA's Research Reforms Fly?")

We had been meeting within walking distance of the Capitol, so we also visited many congressional offices to voice our thanks for the $10 million we were setting out to spend. Of course, thanks were left without regard to whether they had been earned in that particular office, since the visits also allowed us to argue for the importance of the $25 million in the FY 1997 budget just coming under consideration. Marcia Rieke, our outreach coordinator, had managed to arrange meetings for one or more members of the Science Working Group with staff assistants for nearly 10 percent of the total membership of Congress. She had organized a similar effort the previous September. It had been very stressful, with tough questions about SIRTF's cost and worth, and sometimes gruff statements that none of its virtues mattered because the country could no longer afford such projects—our time had gone before it had come.

Later that summer, we made yet another round of visits, urgently seeking funding in the FY97 budget. As with our previous two rounds I drew Tim Peterson, who was the lead science staff member of the House HUD/Independent Agencies subcommittee, from which we needed a positive recommendation. When I had made the first visit, I felt that I had drawn the short straw, since this committee had initially recommended against our getting the $10 million for Phase B. In addition, Peterson was new in his position, not someone who had spent years waiting for a Republican takeover that would move him to the top. No one knew what to expect from him. Indeed, that visit had been unsettling. The small group of us designated for Peterson had wandered through the south wing of the Capitol until we found the committee room, where we knocked and were admitted. We took seats opposite Peterson at the long, polished wooden table running down the center of the room and talked about our dreams for SIRTF. The message seemed to be received in a proper but not particularly warm fashion, and as I left I had no idea where SIRTF stood with him. The reception in May had been a bit warmer, but we still sat formally on opposite sides of the committee table. Peterson had been curious about the scrap of far infrared focal plane I brought along as an icebreaker and as an example of unique SIRTF technology. However, his manner remained properly opaque, and we speculated on what he had been thinking as we walked to our next appointment. This third time I was alone. Peterson was sitting by a fireplace at one end of the room, and he invited me to join him. I started my carefully rehearsed pitch for SIRTF, but he interrupted me after about three sentences. He said that a hard battle had been fought in the previous budget about starting SIRTF, but SIRTF had won and he was opposed to reconsidering decisions already made. I must have visibly gaped, because he hurriedly added that not all the committee members were in agreement with him, but he was working on the recalcitrant ones. He expected to persuade them also, and I should consider the funding for SIRTF to be secure. I thanked him. My business was finished, but it seemed rude to get up and leave. So, I asked him about the paintings on the ceiling and upper walls of the room, which in some ways were quaintly reminiscent of the "heroic worker" posters that came out of the Soviet Union in the 1950s and 1960s. He said they had been commissioned in the 1930s as a WPA project. Even though the paintings appeared dated, they had been preserved because they were so well done and were important historically. We continued to discuss them while the fire crackled, until my appointment had run its course.

While we met in Washington, Simmons had a selection board toiling in Pasadena to pick the industrial contractors in time to present them to the PNAR board. Now, on June 26, 1996, their representatives were sitting in the back of the conference room. Ball Aerospace was going to build the SIRTF telescope and cooling apparatus, and Lockheed Martin would provide the spacecraft and would integrate all the components into a satellite. The contractor selections had just been approved at NASA Headquarters, and the contractors had had to buy their own airplane tickets to the review on speculation that this approval would occur; hence, they would only be spectators. All the PNAR presentations were to be by scientists and engineers from JPL and from the instrument teams.

Other than the lack of participation by the contractors, the review followed the standard NASA format. The presenters sat in the middle of the conference room, in front of the newly selected contractors. The review panel sat around a U-shaped table in the front of the room. The past directors of JPL stared down from paintings and photographs hanging in a formal row on a wood-paneled wall along one side of the room. Inside the U were two viewgraph machines, and the presenters in turn had to descend into this psychological well to be interrogated. Although each speaker had a prepared set of viewgraphs as a framework, the progress through them was frequently interrupted by polite but probing questions. The answers were seldom met with open approval, but if the presenter was lucky, he was allowed to proceed.

Almost before we got started, one of the panel members complained that we were wasting his time. Since we had just selected contractors, he felt certain that our designs would change from what he was to review. He argued that the whole process should have been postponed until we had a final concept. Dennis Botkin, assigned to the review to assess our costs, reminded the panel that a PNAR was not supposed to review designs. Its purpose was to evaluate whether a preliminary concept was worth advancing to the status of a design that *might* be reviewable. Dennis issued similar reminders repeatedly during the day. A PNAR was a new process in NASA, and panel members were not accustomed to confining their questions to the limited set of issues it was supposed to address. In fact, NASA panels often go far beyond the review guidelines in questioning. Dick Novaria of Ball Aerospace summarized the problem succinctly during a break: "Every review is a CDR." A CDR (Critical Design Review) is the final step in the traditional chain that leads to permission to construct hardware for spaceflight, and it is expected to delve into any minutiae of the design that the panel members find of concern. Our CDR was not supposed to occur for more than two years.

As the review proceeded, an appropriate member of the SIRTF team represented each project element or engineering discipline in a half-hour presentation. We had rehearsed the previous week in a meeting tinged with the nervous anticipation of really starting the project. Today, the talks all went smoothly, and the interruptions and questions posed by the panel were taken easily in stride.

Toward the end of the session, however, this flow was disturbed by requests for detailed schedules. The instrument teams had all prepared schedules that would carry us through at least the Phase B study. But because the industrial contractors had had no opportunity to participate, the project itself had only a very high level and nominal plan. We unpersuasively tried to do the NASA review equivalent of discussing the weather, but then an even more worrisome problem emerged. We were proposing a version of the project that came close to the required total budget level of $450 million up to launch and $700 million including launch and operations, but *we did not match up with the annual allocations*. This was not entirely our fault. Administrator Goldin had demanded a less than four-year development cycle, and NASA Headquarters had promoted a launch for SIRTF in 2001. However, the yearly allocations did not provide the total budget for the hardware until well after it had been launched into space on this schedule. On that discordant note, we all broke for dinner at Rich Miller's house. Rich was responsible for the required data reductions and the interface with the astronomy community. He is also an avid model railroad hobbyist, and on that beautiful California evening we snacked in his backyard while miniature trains chugged around the borders.

I found it difficult to enjoy myself and hitched a ride back to my hotel with Art Fuchs, chair of the review board. To my relief, he revealed that he thought the review was going reasonably well. What about the budget issue? I asked. Art replied that the most effective managers had a knack for presenting the problems they couldn't solve to review boards in a manner that led the board to take up their cause. He said that he and Simmons had even discussed this possibility in the weeks prior to the review.

The next day, Simmons presented some very general schedules, and the board wrangled further about the budget. Botkin seemed to smooth over the issue by arguing that the review was designed to advance us into Phase B, we had an adequate budget for that step, and the problems for later project phases were not of concern to the PNAR. He felt that a plan that reconciled the budget and schedule would be a requirement for the Non-advocate Review (NAR) that would judge our readiness to advance beyond Phase B into detailed design. The NAR would occur toward the end of Phase B, not at its beginning.

The next day, June 28, Mike Werner informed the Science Working Group that the review board had given us passing marks on all counts and would forward a recommendation that we advance into Phase B to NASA Headquarters within a month.

On August 30 I received an e-mail message from Mike: "PLEASE NOTE — URGENT. We will have a special telecon to discuss a hitch in the SIRTF schedule (potential ~ 1 year launch delay) on Tuesday, 9/3. . . . I realize the time may not be optimum for everyone, but it is the best compromise I could cobble together." I read Mike's e-mail a couple of times, with growing dread. It certainly confirmed the little indications I had been picking up since June that things were not settled after all.

When we were all connected to the teleconference, Simmons explained:

> Everyone at Headquarters is in support of SIRTF. However, the Office of Space Sciences has to go to the Program Management Council to advocate a project moving forward. Evidently no one has ever done this with a PNAR on the record—we are following a policy never implemented before. The OSS [Office of Space Science] management realized they needed to be briefed in advance to see that we were in alignment with their program. The PNAR and NASA independent cost estimators had endorsed the plan for a launch in 2001. OSS hadn't provided an adequate funding profile for this launch. In response,

> the PNAR panel had recommended that NASA find the funding, and the independent cost estimators said either do that or slip the start of the project so a stretchout wouldn't threaten an overrun. I'm trying to convince the independent cost estimators to agree to let us start now [and stretch out the program], since it seems OSS really doesn't have the money and NASA seems unwilling to help from elsewhere in their budget.

While we all looked for ways to avoid unnecessary financial commitments, Simmons advanced a new proposal to headquarters. We would handle the necessary stretch in schedule by building SIRTF in pieces, none of which would be allowed to expand the duration of its construction from that required for the 2001 launch. The instruments would be built first, then the telescope and dewar, then the spacecraft. There would be some extra costs, because project elements not in construction would have to monitor those that were proceeding so that in the end everything would fit together. However, each cost estimate should hold approximately. The critical underlying assumption was based on the proposal by Lockheed Martin to provide a spacecraft along the lines of those it was developing for two missions to Mars (and for other uses as well). With most of the hardware already proven, they could postpone building ours and also eliminate the unexpected problems that arise on the first build of any new design. Furthermore, the Mars missions were using the same computer we would use, so Lockheed planned to reuse that software, shortening the software development schedule to be compatible with Simmons's program.

We would find out if this approach would convince NASA to let us proceed at the next meeting of the Program Management Council on September 19. Simmons was apprehensive because Botkin, who had turned out to be a strong supporter, would be on vacation. Any further delay would push the approval into the next fiscal year. Given how hard we had had to work to get the funding, it would be very risky not to commit it in the year for which it had been granted.

> Space Infrared Telescope Facility Takes a Step Forward
>
> NASA's Space Infrared Telescope Facility (SIRTF) has moved one step closer toward its journey into space to explore the birth and evolution of the universe.
>
> The high priority astrophysics mission has successfully completed

its preliminary analysis phase—known in the aerospace industry as Phase A—and now begins its definition phase, known as Phase B. NASA granted the approval after an independent review board appointed by the agency found the SIRTF mission's scientific objectives are achievable with the available resources. In this next phase, the mission's preliminary design will be developed.

"This is a prime example of NASA's 'faster, better, cheaper' approach to space exploration," said SIRTF Project Manager Larry Simmons. "Through innovation and new technology, we've reduced the cost while still providing the performance of earlier telescopes. The National Academy of Sciences has identified SIRTF as the highest-priority major U.S. astronomy mission for the 1990s."

In addition to its role in the Great Observatories Program, SIRTF also marks the first major step in NASA's Origins Program, a series of missions designed to study the formation and evolution of galaxies, stars, planets, and the entire universe.

Some of SIRTF's innovations include a unique solar orbit (trailing the Earth as it moves around the Sun), state-of-the-art infrared technology, a new, lightweight cryogenic telescope made entirely of beryllium, and a cost-saving telescope cooling system that reduces the amount of cryogen used to maintain the low temperatures needed for sensitive infrared observations. NASA will request approval to begin SIRTF's design and development (Phases C/D) in fiscal year 1998. (JPL press release, November 12, 1996)

Simmons's strategy had succeeded. Earle Huckins (deputy associate administrator for Space Science) had intercepted the PNAR report endorsing our start but noting that Space Science had not provided the necessary funding profile. To avoid taking this position to the Program Management Council (PMC), he initiated further negotiations within Space Science that had solved the problem in time for the September PMC meeting.

8

SUCCESS BREEDS SUCCESS

■ Although we were proceeding with Phase B, the cost and necessary commitment from NASA to build SIRTF would go up substantially in Phases C and D. We still needed a little luck to get approved, and that luck would have to come from successes elsewhere in Space Science.

> Night Launch Draws Spectacular Reviews
>
> Like a time-lapse sunrise in the middle of the night, Shuttle *Discovery* turned a launch into a fiery extravaganza for residents and tourists up and down the Brevard coast. In the glare of the rocket's ascent, the open mouths and wide eyes of spectators said more than their words could.
>
> The flight got under way on the mark at 3:55 A.M. as *Discovery* bolted from its Kennedy Space Center launch pad, arching over the Atlantic Ocean on a tail of fire that cut through the darkness, . . . setting the stage for four highly complex spacewalks to install new instruments to enhance Hubble's vision.
>
> After the crew reached orbit, they quickly got down to work on a $795 million mission as tricky as any NASA has flown in 82 shuttle flights. The astronauts will aim to tune up Hubble without accidentally

> damaging the observatory, which has been redefining humankind's understanding of the universe since its blurry vision was fixed by space-walking astronauts in 1993.
>
> "Hubble has become a superstar for space science and for astronomy around the world," said Wesley Huntress, Associate Administrator of Space Science with NASA Headquarters in Washington, D.C. (*Florida Today*, February 12, 1997)

Indeed. The countdown to launch of STS-82 had proceeded flawlessly, unlike typical space shuttle launches, which suffer delays due to unsuitable weather or to some problem in one of the millions of parts that must all work to ensure success. The spectacular event was witnessed not only by residents and tourists, but also by hundreds of astronomers (including me) and engineers whose careers had embraced parts of the mission. Others attended just to be present—and seen—at a historic event.

The VIP launch viewing area was a set of bleachers 3.6 miles from the launch pad, across a body of water. The crowd waiting in the cool, damp early morning seemed festive; it was an opportunity to greet old friends and meet new ones. Some checked the status of launch preparations through a small telescope. A TV set up in front of the bleachers broadcast launch details over NASA Select, but it received perfunctory attention at best.

Space Science's luck had certainly turned, at least for the moment. The countdown went so smoothly that the ten-minute holds placed in the sequence as contingencies to provide time to correct minor problems in fact just marked time for ten minutes. The shuttle lifted off on a spectacular column of fire precisely at the earliest possible moment that allowed rendezvous with the Hubble Space Telescope. The roar of the engines reverberated through our bodies and engulfed our senses. The column became a flaming tail, then a glowing star, fading hundreds of miles downrange.

The setbacks to Space Science in the early 1990s had been fixed or forgotten. One of the early targets of the restored Hubble vision had been the sites of the crash of Comet Shoemaker Levy 9 on Jupiter. This event had riveted the attention of professional astronomers and the public on the role of similar incidents in the past and future of Earth. It was rare for professional astronomers to be engrossed in a topic that could be observed and appreciated with any amateur astronomer's telescope.

NASA administrator Goldin had directed that NASA's future would concentrate on a search for planetary systems capable of sustaining life. Huntress had seized the opportunity and persuaded him to establish an "Origins" program that had captured the fancy of the public and of Congress almost before the directives left Goldin's office (Huntress 2003). Tiny wormlike formations, similar in appearance to fossils of primitive bacteria from ancient Earth strata, had just been found in a meteor that had been chipped off Mars in a collision with an asteroid billions of years ago and had eventually fallen onto the Antarctic ice.[1] The evidence that some form of primitive life might have developed on our neighboring planet fueled the interest of the public and politicians alike in Goldin's proposal.

A "Space Summit" series of meetings was called to consider ways to increase public spending on space science. Predictably, the result was a strong push toward implementing Goldin's Origins program. Fortunately, SIRTF would demonstrate many of the technologies that more ambitious missions proposed for Origins would require, giving us an important role in preparing for those future missions. Furthermore, the SIRTF science was well aligned with the new theme. For example, although in general planets are very hard to discover around even the nearest stars, debris disks are a signature of planetary systems that is readily detectable in the far infrared. The study of this emission had already been highlighted as one of the four top scientific priorities for SIRTF observations. Our other core programs on formation of active nuclei in ultraluminous galaxies and on early star-forming galaxies were also important components of the Origins theme.

Ed Weiler had been named to head the new Origins subdivision of NASA. SIRTF had been moved into this area and had been announced as its premier current mission.

9

1996–1997: THE JUNE DEAL

■ All luck aside, if we were to proceed, we had to show that our needs were compatible with the schedule and money guidelines from NASA Headquarters. Simmons's proposal to slide the purchase of our spacecraft to later in the program had only established a basic strategy. Ball Aerospace and Lockheed Martin, who would build most of the observatory, had only barely been selected when this approach was proposed. Once they drew up proper schedules, they would undoubtedly discover unanticipated tasks requiring additional time and money. Getting approval to proceed had already delayed us three months. In the Phase B study we needed to assemble a concrete plan to build SIRTF within the guidelines. And we needed to do it quickly and efficiently, or we would immediately fall off our schedule.

JPL engineer Jim Fanson assembled a System Design Team to reconsider all aspects of the SIRTF design. Each project element would send representatives to JPL for six months starting in July 1996. This period would end with a firmly defined SIRTF to which every project element felt allegiance. The study would also determine interfaces and responsibilities among project elements so each could conduct further design and development without interfering with the decisions being made by the others. In addition, the "co-located phase" was supposed to bond us into a "badgeless" team that would work cooperatively on the project.

The study had started fitfully. Although JPL was providing new facilities for the fledgling project, those for the System Design Team proceeded too slowly to support the planned co-location. To Fanson's frustration, the remodeled quarters were not usable until November, and work on them continued through January. By then, habits were firmly established. The design team had formed a series of "integrated product teams" to resolve the critical issues. The integrated product team chairs preferred to hold meetings down the hallways from their own offices rather than abandoning their families and homes and traveling to JPL. The "co-located team" at JPL became increasingly management focused, primarily in the persons of Frank Martin from Lockheed Martin and Dick Hopkins from Ball Aerospace (plus the project personnel). Fortunately, no overriding flaws were found with the concept developed by the Science Working Group and JPL. The originally envisioned intense period of technical negotiation and ferment proved unnecessary. The bonding also appeared to be proceeding well. Simmons's teaming philosophy allowed the excitement in building a new mission to be shared freely because there was to be no competition to reduce the size of the group at the end of the study.

We were expected to live within a total cost—for design, construction, launch, and operation—of about $700 million, below the cost of the HST refurbishment mission that had occurred earlier in the year. That mission had provided improvements for an existing telescope—extending its capabilities by a factor of two into the infrared and upgrading its visible and ultraviolet instrumentation. The disparity—build and operate an entire mission for the cost of a single major maintenance event in a previous mission in the same series—was a dramatic indication of how much NASA had changed.

Entering the system design phase, the sum of the estimated costs for SIRTF exceeded the cost cap by about 15 percent. Although everyone had been conscious of cost in the engineering studies, there was no indication that any money had been saved. Traditionally, when costs are a problem with a NASA mission, the project managers direct the scientists to identify aspects of the mission performance that can be relaxed. Having gone through multiple grueling cycles of descoping to save SIRTF, however, the Science Working Group was not in the mood to give up additional capabilities. SIRTF was supposed be a Great Observatory for the infrared in parallel with the role of

HST in the ultraviolet and visible. The consensus was that the mission had the absolute minimum complement of features to be viable for this role, and Simmons applied no pressure for further trimming.

Instead, he decided to save money by cramming our activities into a shorter time span. Tight schedules force everyone to work efficiently. In addition, by having more activities occur in parallel to meet an accelerated schedule, some of the fixed overhead costs of just having a project would not have to be paid for so long. Simmons therefore set out to correct the funding profile for SIRTF so that it could be launched as originally proposed, in late 2001. In early February, the release of the administration budget for 1998 demonstrated that he had succeeded.

The rule that faster schedules save money obviously has limitations: nine women cannot produce a baby in a single month. Simmons's success was greeted with dismay by the instrument teams. They would have only eighteen months after final design approval to purchase all the instrument parts, assemble them, develop software, test, correct any errors, and prepare all the documentation to complete their instruments. The instruments' capabilities would depend on sophisticated optical components at the frontier of our technology that had to be special-ordered and produced in minuscule numbers. It typically took a year from placement of an order to the delivery of the part. Once these components had been assembled, the resulting instruments would have to be tested, and that would involve sealing the instruments inside elaborate dewars where they could be cooled nearly to absolute zero. Making the most insignificant adjustment would take weeks—time to allow for warming up the dewar, opening it, and then sealing it and cooling it again. With even a few such cycles, we could expect the instrument testing to extend to five or six months. Simmons's schedule placed the instruments between the rock of critical parts deliveries and the hard place of the necessary iterations during testing.

It was also worrisome to the telescope team, who felt that the low-temperature operation of SIRTF was unexplored engineering terrain. "Engineering has historically focused on room-temperature environments," team manager Tim Kelly explained. "Over the course of only one lifetime, this has changed. [Items like] helium dewars open our design envelope. Engineering intuition, honed over thousands of years, is virtually worthless."

The period from February through May 1997 would be critical to the ultimate success of SIRTF. We had to reduce our costs to match the cap imposed by NASA Headquarters, but we also had to figure out how to build our instruments faster than we believed was possible and to deal with daunting new engineering challenges in the telescope and dewar.

Simmons directed each project element to reestimate its budget requirements. Then the entire SIRTF team would hold a retreat at the Breckinridge Resort in Colorado to negotiate a consistent set of budgets. In the process we were to reduce the total cost estimate to no more than $350 million, leaving $100 million as a contingency fund to deal with unexpected problems. The remainder of the $700 million mission cost was for a rocket and for operation of SIRTF after it was complete.

Setting aside $100 million for a rainy day may seem extravagant to those who do not understand the meaning of NASA cost estimates. It is occasionally possible to get an aerospace company to bid to do a new program at a "fixed price," meaning just what any normal human being would think a price would mean—the item is to be delivered for the specified sum. To avoid any possibility of losing money for the firm, however, a fixed-price bid has to be very conservative, and for a mission as complex as SIRTF the cost would have been well beyond our means. Instead, everyone was preparing "cost-plus" estimates (the word "fee" after "plus" is traditionally dropped in this nomenclature). NASA would reimburse us for the actual costs of the work we did, plus—if "we" meant a profit-making aerospace firm—an additional "fee" for profit. The project would be obligated to pay for any unexpected problems out of the $100 million contingency fund.

In preparation for the Breckinridge retreat, there was a subtle change in the budgeting rules. Previously, we had estimated the most likely cost of each mission element, offering a number we felt had roughly equal chances of being above or below the final cost, and hence termed a "fifty/fifty" value. The tendency for estimates to be optimistic was so prevalent in the early stages of a project that the ensuing cost increases could not be classified as unexpected problems. The project contingency—that $100 million—was not to be used to cover normal cost growth. Therefore, Simmons had attempted to redefine the original estimates as "not-to-exceed" values partway through their determination. To emphasize how serious he was on this point, he repeatedly stated that it was his intent to return the entire $100 million to NASA at the end of the project. Unfortunately, this intention appeared to have little effect on the numbers. Therefore, we were instructed to make two estimates: the fifty/fifty value and the not-to-exceed value for which we "promised" to deliver our part of the mission. Work proceeded feverishly through March and April.

It is not easy to predict the cost of a complex NASA mission using an untried management approach to launch new technologies into space five years in the future. Two approaches are possible. One is to construct a "cost model." That approach starts with a cost based on fundamental parameters such as weight, with correction factors for the overall complexity and for the overhead costs associated with the management approach. The entire lot of educated guesses is fed into a "cost-modeling" computer program that inscrutably returns a number. If any of the complexity, newness, management, and risk parameters are adjusted, a new number will dutifully be returned. Some degree of such fiddling is required, since the computer otherwise occasionally seriously misinterprets the inputs. However, the ability to influence the result in this way can make the output of a cost model suspect, and some independent check is needed to confirm the estimate.

Cost models are termed "top-down" estimating: one looks at the final product and compares it with other final products, ignoring the details of how any of them got built. The alternative is a "bottom-up" or "grassroots" estimate. A grassroots estimate starts with each engineer looking at her specific responsibility and estimating how much it will cost to carry it out. The numbers are then added up, possibly with a small allotment for a management reserve, to produce a total for all of the engineering tasks. The design of SIRTF was not sufficiently detailed to make such an engineering assessment of cost very accurate. Nonetheless, Simmons directed us to make grassroots estimates as the only check we had against the parameter adjustments in the cost models.

The results of these estimates and the underlying technical assumptions were presented in a series of reviews in April. It was already obvious that Simmons needed everyone to sign on for the fifty/fifty budget value to have a chance of staying within our allocation. However, the project would quickly become unmanageable if every element came back for more money as soon as it found a problem. Thus, the fifty/fifty budgets needed to be viewed as "not-to-exceed." Not surprisingly, the data presented at the reviews failed to solve this dilemma. Everyone understood that they would have to live with whatever budget they proposed, and the estimates were thus substantially higher than previously.

An e-mail message from Simmons went out to the team leaders on April 18, 1997:

> Management Team:
>
> Based on information received at the [reviews] that have been completed to date, it is clear that the budget situation cannot be resolved by simple adjustments. Therefore, the SIRTF Work$hop which was scheduled for May 20–22 in Breckinridge, CO, will be canceled.
>
> Each system area will be contacted next week to discuss actions required to resolve the current budget situation. While a plan has not yet been formulated, each system area should be giving consideration to what changes would be needed to restore the budget for their areas to the levels presented at the SIRTF Preliminary Non-advocate Review in June of 1996.

Our effort to hold a firm line on science capabilities and still stay within NASA's hard cost cap had failed. At the same time, our options to deal with the problem had been narrowed at a Science Working Group meeting in Washington at the end of April. Wes Huntress met with us and told us how excited he was at the prospects for Space Science:

> The last five years have been pretty tough, but last year was a banner one—we started over a dozen new missions. We had to restructure to smaller missions, tighten construction schedules. The productivity of Space Science is very high, plus it has immense public appeal—you can't read a paper for more than a week without finding a Space Science–related story. A lot of things happened this summer—the Mars rock was a precipitating event. It highlighted "Origins" in a way that the public can personalize the program. They begin to understand life cycles in the Universe—it presents to them that science is a drama.

He had negotiated a matching arrangement with Goldin to put additional money into Space Science as Huntress found ways to save on the operations of existing missions and on administrative costs, and the FY98 president's budget showed full funding of the $450 million allocated for SIRTF.

Ed Weiler, as head of the Origins program, attended the entire meeting. After Huntress left, Simmons described our current difficulties. We assumed he was hinting that we would have to take the usual route of reducing the content of the mission. There was an uneasy pause while we mentally mustered all our well-worn arguments about how further cutting was impossible. Instead, Weiler broke the silence, pointedly stating: "Mr. Huntress and

Mr. Goldin have solved the funding profile problem and found the [$450 million], and they would be very upset if you proposed to descope the project." Another pause followed, even longer than the first. Simmons finally said he would begin looking for ways to improve efficiency to bring the budget into line at our next management meeting on May 8.

As we gathered at JPL that day, the mood was down. To get within the $450 million cap, Simmons was now blatantly substituting our fifty/fifty numbers for the not-to-exceed ones. No one wanted to speak first, for fear of bringing unwanted attention. A turning point came when Frank Martin broke out in his North Carolina drawl: "I don't see why you are all so upset. If I add up all the not-to-exceed numbers, I get $510 million. Although that's more than $450 million, it isn't 15 percent more, which is where we would have a cancellation review. So everyone has said that they are confident they can do their part of SIRTF, and we wouldn't exceed our budget by enough to be threatened with cancellation." Martin's numbers left out the $100 million project contingency fund, so one could have argued that we were about $160 million too high rather than $60 million, but no one brought up that detail.

A plan evolved. Simmons issued "budget guidelines" which in practice quickly became firm allocations, although he argued against that interpretation. By necessity, the targets were close to the fifty/fifty budget figures. Each system put together a budget to meet Simmons's guideline, and each met with him at JPL to discuss it sometime during the next month. These individual meetings provided nominal opportunities to argue against the guideline, although doing so would have required considerable courage. Once that opportunity had passed, one had accepted the guideline implicitly as a true allocation. Nonetheless, a number of project elements claimed later that they felt they had agreed to do more than was possible with their share of the budget.

We were aiming toward another meeting of the entire management group on June 5. The goal of that meeting was described in another e-mail from Simmons on May 30: "The [following agenda] assumes that each team element has a proposed approach that is consistent with the agreed upon target numbers. It is suggested that each discussion leader prepare a brief summary of the impact in their area, and devote most of their time to give-and-take discussion rather than an extended presentation."

In addition to our struggles with the budget, we needed to find a way to build instruments on the schedule demanded by Simmons's "victory" in moving the launch date back to 2001. A possible solution was suggested. By bending the NASA rules, we could make critical long-lead purchases prior to approval of the overall instrument designs for construction. This approach would increase the demands on the funds available for 1998, when we would have to make many of these purchases, but it was the only way to get the instruments built on time.

The June 5 meeting turned out to be anticlimactic. Simmons summarized our progress in an e-mail on June 9:

> The Team meeting last Thursday reinforced my belief that we can build the SIRTF we have been talking about. The entire team should feel very good that we are very close to having an approach that is consistent with the resources available. We are still working the numbers here in the project office, and there is still a lien against the Telescope plan. But all-in-all we seem to be in pretty good shape.
>
> The challenge will be to move the design forward quickly enough to have a credible PDR [Preliminary Design Review], and at the same time complete the project planning needed to convince the NAR we are ready to go.
>
> Hopefully, we can now focus on planning how we are going to proceed knowing how much we have to spend. This has been a difficult period, but I felt the team was truly trying to find a solution, and we did! Thanks for all of your hard work.

With the exception of the telescope team at Ball Aerospace, which had been particularly hard-hit by Simmons's targets, everyone came to the meeting with budgets that fitted. The telescope team had had to propose eliminating spare optics, putting us at risk for the whole mission if the flight optics were damaged as we assembled the observatory. No one else had significantly reduced their responsibilities for the mission. Thus, over a four-month period we had all agreed implicitly to treat our fifty/fifty cost estimates as if they were not-to-exceed costs, with some minor adjustments in management approach to convince ourselves that the change was acceptable. Our only protection against unanticipated problems, our own errors, and accidents of fate was the $100 million reserve.

We had talked ourselves into a fast development cycle and a much lower budget than had come to be expected for missions of the complexity of SIRTF. We had been waiting twelve years for this chance and could not afford to blow it. Don Strecker, my systems engineer at Ball Aerospace, quipped, "Faster, better, cheaper: pick any two!" but we knew we would have to do a lot better than that.

10

LEARNING TO MANAGE

■ The negotiations leading to the "June deal" had had the potential to pull our team apart. The team had survived, though, partly because we had more or less been equally pressed. In addition, Simmons had insisted that all the negotiations be carried out with open books: everyone knew everyone else's budgets and proposals. Although open books provided some on-the-spot controversy, in the long run the approach was critical in eliminating paranoia. Still, we needed something positive to allow the strains to mend.

Ideally, we needed a group activity focused on management issues, but with a light-hearted twist so that we could put behind the day-to-day details. Charlie Pellerin, formerly head of NASA Astrophysics, was now at the business school of the University of Colorado, just a mile or two from Ball Aerospace in Boulder. Outside the university he ran a management-training institute that specialized in exactly the type of event we needed. In addition, some of us were in serious need of management training. Nearly any reasonably successful person assumes that he or she has learned to manage well. There is some truth to this assumption, since success in most areas requires that one not be completely incompetent as a manager. However, the typical education of a scientist includes nothing about management, yet the three instrument principal investigators now had management responsibilities that demanded far more than a mere lack of incompetence.

We first met with Pellerin on June 26, 1997, in a conference room at Ball surrounded by corridors where engineers were designing two of the SIRTF instruments and the telescope. Pellerin and a colleague distributed a booklet that was to be the focus of the session. The first page, headed "Hubble Space Telescope—How this all began," read:

> In the fall of 1990, I [Charlie Pellerin] walked into the St. Louis airport lounge in high spirits. My week in Japan had been just what I needed. Although the Hubble Telescope was finally in orbit and working well, the path to success had been stressful.
>
> As my thoughts turned toward Washington, I called my office to check in. I soon heard Len Fisk (my boss) saying, "Charlie, what do you know about spherical aberration?" As I wondered why he might be asking, I replied, "I only know that it is a common mistake of amateur mirror grinders."
>
> "Well, what would you say if I told you Hubble has it," he asked? I answered with the first thought that entered my mind. "I would say that you are annoyed that I had a good time in Japan, while you had to tend to the bureaucracy."
>
> He persisted, but I remained convinced that he was joking. He then said, "OK, look around the room and bring the front page of any major newspaper back to the phone." I did so, and he directed me to read the headline over the phone. It said something like "Hubble Mirror Flawed, a National Disaster." "Now what do you say," he asked? I replied, "This is really something. How did you plant a fake newspaper in here?"
>
> Reality sank in the next day, back in Washington. A trivial and obvious error overshadowed the accomplishments of thousands of dedicated people! Life would never be the same. The Congress, pundits, and even late night comedians denigrated our team.
>
> NASA formed a Failure Review Board to find out what had happened. The Board discovered that a technician working alone made a subtle error in adjusting the device used to measure the mirror's surface. At first we felt a sense of relief. It wasn't management's fault. We then realized that no project could be so vulnerable to one person's mistake. Our complex NASA quality assurance systems are intended to prevent just such occurrences.

> Then, a more disturbing realization—there had been hints of a problem that were not acted upon. How could this happen? The review board finally concluded that the failure was a *leadership failure*!
>
> From that day forward, my passion has been to understand, and then teach, organizational leadership. . . . My life purpose is now to be a teacher and leader of leaders.

Pellerin's message was deceptively simple and simply presented. His definition of leadership was "influencing others to do willingly and do well, that which has to be done." The course used simple diagrams and cartoons. The first drawing represented four essential components of leadership: team leadership, inspiring leadership, directive leadership, and visioning leadership (Figure 10.1).

Good leadership was described as drawing on all four components while remaining at the center. In that way, a leader can plan in groups and encourage a team spirit (upper left), relate to personnel's needs and goals (lower left), direct underlings to do what is needed (upper right), and create concepts that will push the effort to a new level (lower right) as necessary. Pellerin illustrated many of his points with examples from NASA successes and failures, including the problem with the Hubble mirror.

For breaks, we went out into the parking lot in front of the building and did group calisthenics. If the meetings had no other benefit, the parking lot sessions kept the Ball engineers who watched these activities from their offices thoroughly amused, as they enjoyed reminding us for years.

"TEAM LEADERSHIP" Relating to create trust and build synergistic teams	"DIRECTIVE LEADERSHIP" Organizing and directing people and resources
"INSPIRING LEADERSHIP" Giving meaning to people's work and lives	"VISIONING LEADERSHIP" Creating concepts which pull followers to a desired future

Figure 10.1. Four Styles of Management. A good leader uses all four approaches to leadership.

11 PRELIMINARY DESIGN REVIEW

■ Although NASA prefers to use the Phase A–D nomenclature, a space project goes through characteristic emotional stages as well: euphoria (initial selection), optimism (original definition), confidence (design and initial construction phases), concern (as problems emerge in construction), panic (as problems grow), confidence (problems solved), and bliss (successful launch) or despair (unsuccessful). Although our long wait to get going had imposed a number of out-of-sequence concern-panic cycles, we were now on the usual track. We had just emerged from the optimistic phase and were entering the confident phase of design and initial construction.

As a result of the budget negotiations, each SIRTF system team now had a clear understanding of what had to be done, on what schedule, and with what resources. The System Design Team's activities had established a projectwide framework to combine our efforts. The summer of 1997 was spent polishing up the designs accordingly. In August and September we had a series of Preliminary Design Reviews (PDRs) for each system. In a PDR, one is expected to show an understanding of the requirements and to present a design in sufficient detail to show that it can meet those requirements (within the resources available). We decided that these system-level reviews could be relatively informal, encouraging a free exchange of engineering concepts and concerns from the reviewees and ideas and critiques from the reviewers. Although such an exchange might be expected as a matter of course in a review, as the degree of formality increases, so does the guard of the parties being reviewed. Very formal reviews therefore can be less informative and searching than informal ones.

The system-level reviews were followed by an all-SIRTF PDR on September 23–25, 1997. To avoid an overly extended review season, Simmons had scheduled the PDR and the NAR (Non-advocate Review) to run concurrently. The NAR is supposed to be carried out by a panel of knowledgeable experts not involved in any way with the project, who are charged with taking a hard look at whether it can live up to its financial and schedule commitments. The PDR panel is charged with oversight of the technical requirements and the design to meet them and is supposed to be helpful rather than judgmental, so the two reviews are complementary.

Between the series of system-level reviews and this projectwide one, more than 10 percent of the project's resources for the year were devoted directly to the series of PDRs. Additional reviews earlier in the year added to this share. Yet, we were also trying to operate in an exceptionally efficient manner with a minimum of project oversight.

Simmons kicked off the presentations with a description of the program background, emphasizing how we had achieved the June deal. He also described the "level-one requirements" for SIRTF—the most important requirements for the mission, basically a compact between NASA Headquarters and the responsible NASA center. Any threat to them is taken very seriously. He was followed by members of the project office at JPL discussing first the overall observatory design (Figure 11.1) and then different aspects of it. Although the review was sober and serious, given all we had gone through to get to it, there was a quietly festive air in the back of the room.

The status of the telescope worried the reviewers. Beryllium had been selected as the material for the mirror because it has a very large stiffness-to-weight ratio—about seven times that of glass—and hence could provide a very lightweight mirror. Its disadvantages are many. Its dust is poisonous, so only a few optics manufacturers are prepared to work with it; it changes dimensions significantly upon cooling; and it has internal stresses that distort the shape of its surface when it is cold compared with when it is warm (as it is in the optical shop when the surface shape is established). To deal with these concerns, JPL had Hughes Danbury Optical Systems manufacture a beryllium primary mirror to virtually the same specifications as expected for the SIRTF telescope and had then tested it at JPL.

Normally, that demonstration would have been followed by an order for a similar unit for the flight telescope incorporating all the improvements suggested by the experience with the demonstration one. The flight version would also be subjected to tighter production controls, including documentation of every step in its manufacture (what NASA likes to call the "paper trail"). However, because of the budget squeezing, it had been decided that we would use the demonstration mirror for the flight telescope.

In manufacturing the mirror the Hughes technicians had let too much time elapse between checks of the shape of the surface. To their shock, during this interval the outer zone of the mirror had become raised by about 2 microns from the shape it should have had. As a technology demonstration with no strict performance requirements this deficiency was acceptable. As a flight article it was not. The outer zone needed to be fixed or we would have to compromise a level one requirement. Hughes had proposed fixing the problem by taking a very small polishing tool and working the edge for seventeen hundred hours—a year of normal shift operation—not counting time for testing. The proposal indicated that Hughes was near the limit of its ability to deal with the "turned-up edge."

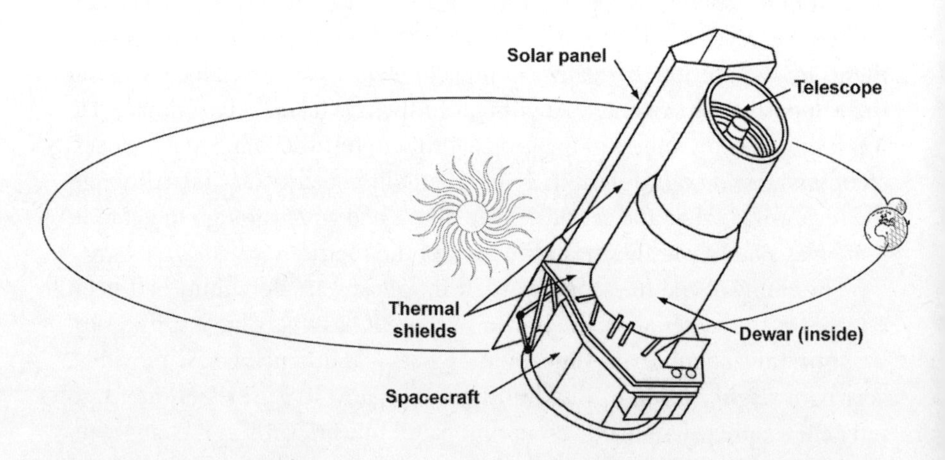

Figure 11.1. The Concept for the SIRTF Observatory. The observatory is in its trailing orbit behind the earth, with the solar panels facing the sun. The spacecraft is at the bottom, and thermal shields protect the telescope from its heat and that of the solar panel. The dewar is inside the telescope outer shell above the spacecraft shield.

Ball Aerospace was preparing a plan to deal with this situation but was not quite ready to make it public. Thus, about all that was offered to the review panel was a high level of confidence on the part of Tim Kelly, the telescope manager, that a satisfactory solution was on the way. On top of the lack of a full plan for the flight mirror, there was no room in the budget to pay for a spare mirror, so anything that went wrong would threaten the viability of the entire project.

The reviewers also worried about our ability to stay within budget. We had presented a list of "descopes"—things we could do to save money if we saw that the costs were getting out of control. Such a list was a mandatory part of the mission management strategy. Unfortunately, almost all of the possible descopes would require decisions very soon, in the confident stage of development. It would not become apparent that the changes were necessary until the concerned and panicked stages, too late to implement them. Eventually it was decided that the only meaningful security against overrunning the budget was the $100 million reserve.

The PDR and NAR are critical parts of the NASA rite of passage from Phase B (conceptual design) to Phase C (detailed design), and all the reviewers felt that SIRTF had qualified for this step easily. Official confirmation would require a briefing to the Office of Space Science on October 24, followed by a presentation on October 29 to the Program Management Council, which would draft a "Program Commitment Agreement" that would be sent to the NASA administrator, Dan Goldin, for his signature. With the strong endorsement of the review board, we expected this process to go smoothly. Equally important was the news provided at our mid-October Science Working Group meeting by Ed Weiler: "The 1998 budget is final, and it is solid for us. SIRTF is a New Start. . . . The outlook after 1998 looks good, but we don't know what we will do about Space Station *Freedom* overruns. However, once you get a New Start, you are usually OK—there is a tendency to cut things that aren't started yet." Simmons pointed out that the budget included not only the minimum needed for the Phase C work, but also $14 million to establish a reserve and for the long-lead purchases we were depending on to hold to our schedule.

At our quarterly review on November 12, our program executive at NASA Headquarters, Lia LaPiana, reported that the "presentation to the Program Management Council resulted in a recommendation that we proceed, but final approval was withheld pending submission of a risk management plan and a definition of mission success criteria." The delay created a potential funding problem because we were coming to the end of the Phase B funds but could not spend Phases C and D funds until we were formally approved. We needed to submit the two requested items before the next meeting of the Management Council on February 22.

In the January 22 management meeting, however, we learned that the presentation to the Office of Space Science on October 24 had been received almost like a new review. A surprisingly long list of requests for new activities had been sent back to us in mid-January. The OSS wanted answers to such technical questions as how we would ensure alignment of the telescope and instruments and the reliability of the aperture door mechanism that had to open to let light in to the instruments; how we would deal with the varying helium usage by the instruments; and how radiation would affect the electronics components. On the management side, there was more concern about the lack of a viable descope plan, how the teaming would work when times got tough, and the mission success criteria. We were also to provide additional details about safety and mission assurance plans, cost information, identification of the critical path (a management term for the activities that have the greatest schedule pressure), schedule risk due to the telescope mirror situation, and identification of schedule reserves.

In frustration, Jim Houck asked: "Do we say we have a New Start?" Simmons directed us to assume we could get one by concentrating on getting approval by the Program Management Council and then to go back to Space Science and deal with the new issues they had raised. However, he also urged us to overlook the details. Congress had approved SIRTF for a New Start in FY98. NASA had identified only minor items that needed to be clarified or completed. Goldin had stated at the American Astronomical Society convention in early January that SIRTF would have a New Start in FY98. Huntress had echoed: "Everything is in place for a New Start for SIRTF, and we'll be going ahead with it."

The presentation to the Program Management Council on February 22 was successful. At the March management meeting Simmons reported: "The Program Commitment Agreement is to be signed March 25, which will be defined as the official New Start. The CDR will be in September with the IAR [Independent Annual Review] about a month later. We will have a round of systems-level reviews before the CDR to be sure we have gone into depth. We are to encourage members of the CDR board to attend the systems level reviews so they have more insight to the program."

On March 25, 1998, JPL released the good news to the public:

NASA Administrator Daniel S. Goldin today authorized the start of work on the Space Infrared Telescope Facility (SIRTF), an advanced orbiting observatory that will give astronomers unprecedented views of phenomena in the universe that are invisible to other types of telescopes.

"The Space Infrared Telescope Facility will do for infrared astronomy what the Hubble Space Telescope has done in its unveiling of the visible universe, and it will do it faster, better, and cheaper than its predecessors," said Dr. Wesley Huntress, NASA's associate administrator for space science. "By sensing the heat given off by objects in space, this new observatory will see behind the cosmic curtains of dust particles that obscure much of the visible universe. We will be able to study fetal stars, detect other solar systems, and study the most ancient, distant galaxies at the edge of the universe."

SIRTF, whose design and development is cost-capped at $458 million, will be one of astronomy's most advanced telescopes. Its unconventional approach uses new technologies, an innovative mission design, and small launch vehicle. It is being developed on a quick schedule that closely integrates the work of the contractor and academic teams responsible for development and delivery. Its design promises high sensitivity and observing capability along with efficiency of operations and long lifetime.

The project also represents a bridge to NASA's Origins program, which seeks to answer fundamental questions about the birth and evolution of the universe. SIRTF will lay the groundwork for many investigations fundamental to the Origins program, such as studies of the birth and evolution of galaxies, their stars, and searches for planets that orbit some of those stars.

12
NEW PROBLEMS FOR A NEW MANAGER

■ We were entering the "concerned" phase of development. At this point our designs would confront reality as we tried to fill in the details and build and test them. The level of concern also rises at NASA Headquarters during this phase, requiring a project to solve its problems under the scrutiny of a phalanx of reviewers.

Under the funding that Simmons had reported in mid-October, we started to acquire the long-lead items needed for our instruments. Even with this head start, many of them were delivered just in time for the instrument construction schedule. This critical funding came with an ironic twist. At our February 12, 1998, management meeting, Simmons reported: "It appears the $14 million we got extra was really a way for headquarters to hide money they wanted to spend on the space station." They were going to take the money out of our FY99 budget to pay the space station back, leaving us with a deficit for that year relative to our plan. He added that Space Science was trying hard to find a way to reestablish their commitment to us for FY99. Fortunately, the money was eventually replaced.

The organization of the SIRTF Science Center (SSC) had also been completed by October 1997. It had been decided some time before that the SSC would be built on the capabilities of the Infrared Processing and Analysis

Center (IPAC), which had been established on the California Institute of Technology campus to reduce data from IRAS. IPAC had a reputation for providing good service to the astronomy community on a modest scale that was well suited to the style of SIRTF. However, the SSC needed a director. JPL hoped that the costs could be reduced if the SSC were closely integrated with Cal Tech, and Cal Tech insisted that in that case the director should be selected from its faculty. Fortunately, Tom Soifer was willing to undertake the job. He was not only a professor at Cal Tech and a well-regarded infrared astronomer, but he had also distinguished himself in his management of the IRAS data reductions. George Helou, who had been the acting SSC director, became Soifer's deputy.

Our annoyance at the seemingly endless progression of reviews was exacerbated in April when Program Executive LaPiana announced that she was appointing an External Independent Readiness Review (EIRR) board—one more panel to keep an eye on us. Further, we had to pay the expenses of the EIRR out of our project reserve, a minor but irritating example of how unexpected obligations crop up. One JPL team member bluntly stated at a meeting the next week that "an EIRR is a pain in the butt—it is a whole new set of characters that aren't up to speed that have to be satisfied and can sink the ship because they have a separate line of access to NASA Headquarters." The concept of an EIRR had been created for the first HST servicing mission when the corrective optics were installed. The future of NASA Space Science was considered to ride on the outcome of that mission, justifying the extra cycle of review. Imposing an EIRR for a mission as small as SIRTF suggested an escalation of oversight by NASA Headquarters contrary to its own standards for faster, better, cheaper.

In addition to our exasperation with the EIRR, we had contradictory feelings about our own reviews. On July 22, 1998, we had one of the first systems-level Critical Design Reviews (CDRs), where an argument broke out about a new documentation requirement on the instrument teams. Although the dispute was brief, we felt its repercussions at the next management meeting. Simmons began: "We don't seem to be impressing boards well that we have our acts together." Only after a few minutes of discussion did it become clear that the issue was the argument at the review. "In the future, the system under review should give the presentations and we should leave it to the review board to give feedback," Simmons scolded. This position was too much for some members of the group, who immediately protested that going to listen to two months of reviews and not being allowed to discuss and clear up misunderstandings was both a huge waste of time and potentially detrimental to the success of SIRTF. The debate split along predictable lines. The managers argued that the reviews were "a gate we need to get through" and urged us not to "rock the boat." On the other side were the scientists contending that we needed these forums to exchange information and that the gate we needed to get through was the full facility-level review in September. No conclusion was reached, but behavior in the systems reviews became more circumspect.

We had set ourselves up for this problem. We had wanted the facility-level CDR board to feel well connected to the project and had invited the members to the systems-level reviews so that they would have, if anything, an overdose of insight. By inviting them, however, we had created the temptation to treat the reviews as a show. We had no other channel for private expression of concerns about the project elements (that, after all, is the ostensible purpose of reviews). Thus, we ran the risk of investing a substantial portion of our resources in elaborate public display rather than in achieving the internal insights we needed to detect possible problems in the SIRTF design.

Circumspection was not required for the telescope CDR. When Hughes Danbury Optical Systems had not responded to a request to fix the mirror, Ball had heaved a huge sigh of relief and offered the job to Tinsley Optics. Tinsley had been enticed to accept it on the basis of possible future business making beryllium mirrors for much larger telescopes (Ball and Tinsley became partners in making the optics for the James Webb Space Telescope a few years later). Tinsley had developed advanced methods for working beryllium mirrors and refigured ours beautifully. I thought something outside NASA's starchy review culture was needed to mark the breakthrough and started clapping. The review board joined in.

On June 22 we passed another milestone: the first scientific conference to consider the use of SIRTF. There are many astronomical conferences every year, but this one had the special significance of signaling the emergence of SIRTF, after fourteen years, as a real science opportunity. There were no less than three keynote speakers. Chas Beichman, the director of IPAC, compared the history of SIRTF with "the airplane ride from Hell. You get on the plane, it leaves the gate, and it is brought back for some problem. This happens over and over again—on and off, on and off, on and off. Suddenly you notice it is three quarters of the way down the runway before you expected it to leave the gate, and you have to run after it to catch up." Harley Thronson was more serious, as befitted his position representing NASA Headquarters: "SIRTF serves the highest priority goals of NASA—origins, the search for external planetary systems. . . . Do good work with SIRTF; make it visible even in the infrared." Ed Stone, the director of JPL, extended a welcome on behalf of JPL and Cal Tech: "SIRTF has had a turbulent history, from the shuttle-attached mission to the current one. It is now firmly part of the new era of faster, better, cheaper. It is clear in this era that the key to getting the political support to mount these major missions is to find innovative ways to do them."

The projectwide CDR was held on September 22–24, 1998, in the large von Karman Auditorium at JPL. We had been asked to bring hardware prototypes of what we had designed and proposed to build. These items were arrayed around the back of the auditorium, producing the atmosphere of a technical exposition. The attendees and presenters sat in front of the displays, and the reviewers sat in a horseshoe-shaped arrangement at the front of the hall. Each presenter stepped to a viewgraph machine in the center of the horseshoe to present an overview of his system, a description of its system-level CDR, and a statement about what was to be done about the recommendations from that review.

Our analyses were now in considerable depth, and it was often difficult for reviewers to ask sensible questions on the details. Despite the opportunity to attend the systems CDRs, most of them had not interacted with the project much during the preceding year. Thus, we heard:

- Why do you cantilever the solar panels off the spacecraft and don't attach them to the telescope? [The design tried to minimize the thermal path between the hot solar panels and the telescope shell, which had to run very cold to preserve the liquid helium.]
- How can MIPS be on schedule and under budget? You must have poor long-range planning and will run into problems later because of deferred critical expenditures. [A particularly ironic complaint, given the emphasis of the project on being cost effective and accumulating our own reserves.]
- Do you operate the instruments in the [multiple instrument chamber] before putting them in the cryostat [dewar]? [None of the instruments operates anywhere near properly at room temperature.]

Their associated inefficiencies aside, high-level reviews such as these occasionally turn up something critical; and clever managers can use them to help solve problems that are otherwise intractable. Preparation for a review forces a team to focus on its job and can be valuable in promoting overall progress. Reviews also provide gates that all elements of a project must pass through together, in theory preventing uneven progress that might allow part of a project to threaten the whole by falling too far behind. In addition, the reviews have a ceremonial aspect, marking key steps toward the launch pad.

At the end, as is traditional, each reviewer gave a summary. The spotlight worked its way around the table:

- The review was an excellent job; I'm pleased how the chemistry of the team is coming together. My worries are that the flight software is hanging out badly, particularly fault protection.
- The subsystem CDRs were excellent. The telescope is just wonderful—it was a tough decision and you made the right one. I have a very bad feeling about the Lockheed spacecraft team as a whole—you're not getting the team you deserve.
- This is my ninth review, and I want it to be over.
- I'm impressed with the team, and I'd particularly like to compliment the instrument PIs, who are involved in depth. The spacecraft software is not far enough. Overall an excellent review, and you're ready to go on.
- The instrument schedules are too tight. No slack is not acceptable. You need to call back the instrument deliveries, even though I know the instrument people will not be happy.
- I've never seen such a complex cryogenic system not having to be warmed up before it is ready to go.

We had another review to complete before the season was over. In late October, the Independent Annual Review (IAR) was held in a cramped conference room behind the main cafeteria at JPL. The chair of the board, Larry Caroff, began by indicating some sympathy for the preceding four months of unrelenting reviewing: "You probably think this is the YAR—yet another review." A number of the concerns from the CDR reemerged. A new worry was that we were not spending our financial reserve fast enough. We were advised to try to solve our problems early rather than save money and let them grow bigger. As Dennis Botkin put it: "You're starting to push a bow wave, and there is only one way bow waves change: they grow."

The review board reports were processed through NASA Headquarters much as had happened the previous year (but without the delays and growth of questions). We received permission to advance into Phase D.

In the management meeting immediately following the IAR, we discussed a proposal to accelerate building the warm electronics for the Infrared Spectrograph (IRS) and MIPS by additional automation of testing and quicker purchase of key parts. In exchange for funding these augmentations (at $750,000), Simmons extracted a requirement for IRS to be completed two weeks earlier than previously agreed, and for MIPS to show two weeks' reserve on its completion date.

The change in the instrument schedules was small but symbolic. Some of the other problems had deeper roots. In the November management team meeting Simmons reported: "Software at Lockheed is our biggest problem. LMA [Lockheed Martin Astronautics in Denver] is also in trouble and we depend on them for software transfer. All programs currently being developed in NASA Space Science are in trouble with software."

We devoted the December meeting to further discussions of the spacecraft software issue. The spacecraft manager, Milt Whitten, summarized the problem: "We proposed a total of 45,000 to 53,000 hours of effort, but we now expect to have to invest 134,000 hours. We had assumed when we developed the spacecraft cost that we could easily reuse the LMA software being developed for Mars Climate Orbiter and Mars Polar Lander. However, we have discovered the Denver software is not modular, so it is not easy to reuse. Denver has its own problems and is not offering much help. The requirements are not well defined. . . . That has also held us back seriously. Our cost models suggest a cost impact of about $13 million."

Meanwhile, a new problem had developed. At the CDR, the IRAC team had presented a schedule that included completion of their instrument in June, with nearly a year of testing before delivery. By November they had encountered multiple cost and schedule problems. They were having serious troubles with the suppliers for their optics, particularly with the optical filters. In addition, they announced that they had cut all support out of their budget for operating their instrument while it was being tested in the spacecraft; all they were prepared to do now was to "support someone else adapting their procedures to [the JPL interface]."

In January, the telescope team revealed their reaction to the CDR comments about inadequate schedule reserve. Dick Hopkins reported that Ball Aerospace had "developed a plan for a double shift integration team (at a cost of $1 million) which gives a one-month schedule improvement. This is a direct response to the CDR." This reserve eventually grew to seven weeks after a review of all the proposed tests and elimination of those that appeared not to serve a useful purpose. There was also more optimism about the spacecraft software. Now that Lockheed had delivered the Mars mission software, it was felt that we could get more attention. Simmons was optimistic: "The Lockheed Martin organization has been changed to try to contain overall corporate costs. They are streamlining divisions, allowing for mixing of people between civil and military worlds."

IRAC also seemed to be progressing. Simmons enthused: "Having been to Ball last week and on the East Coast [to Goddard], I conclude we're still in pretty good shape. We have been through the review process and have come out well and are now considered an icon for how to get things done. Let's not get ground down by everyday concerns. We have the money to build it and fly it. So let's have a good time and do that."

However, IRAC brought more problems to his door in February. A high-level manager from Goddard came to negotiate a significant increase in funding, bringing a detailed list of problems that the money was needed to solve. Simmons grumbled: "We agreed to do this as if it were fixed price even though the contracts are cost reimbursable. But now we're falling back to 'I can do that if I get extra money for it' and we're getting constipated in the decision making. It is true that I hold a reserve—but I never told you not to put some money aside to get higher confidence that you could get your job done, and I'd hope you did that."

Given the subsidy IRAC was already getting through civil service personnel at Goddard, for which the team paid only a small fraction of the true cost, the adjustment was also annoying to the other instrument teams who were paying full costs at Ball Aerospace. Tom Roellig, who was on the IRS team (and had taken over as facility scientist when Frank Low had semiretired from the position), asked: "We have squeezed blood out of rocks—are they at the same level of pain?" Bill Irace, who oversaw the instrument construction for the project office, responded: "There's no doubt they need this. Without it they won't deliver."

Simmons tried to soften the impact: "IRAC hasn't participated like the other instruments in design teams and other project activities. This should fix that." But he was immediately undermined by Lois Workman, the IRAC manager at Goddard: "This addresses getting the instrument built. It hasn't addressed coverage for other activities." There really wasn't any choice, however, and after a perfunctory negotiation IRAC was given more than $1 million from the reserve.

By February the project had done what it could to react to the various problems that had been exposed by the series of systems reviews and by the facility review in September. We had also heeded too well Botkin's admonition to spend our reserve more quickly. The spacecraft team was asking for more than $13 million extra to fix the problems in the flight software, plus another $1 million to cover a price increase for a key communications item. Ball was asking for $1 million to create a reserve in their testing program, and another $1 million had gone to IRAC. The costs to accelerate IRS and MIPS had also come to nearly $1 million. Nearly half the project reserve was gone, and we had a long way to go to launch.

Simmons summarized the situation to the Science Working Group on April 5, 1999: "NASA has been very supportive of us, once the project got started. They haven't subjected us to annual replanning, sent us half the money, or anything like that. They've given us what we asked for." Some years before, an agencywide study chaired by former Martin Marietta CEO Norman Augustine had noted that the constant replanning NASA put its missions through in reaction to changes in the funding possibilities was a major contributor to inefficiency (Augustine et al. 1990). Simmons was pointing out that we should not reward NASA for correcting this problem by blithely overrunning our allocation and asking them for more.

It was not a good time to ask for funds in any case. At a Space Science monthly staff meeting at NASA Headquarters on May 24, Ed Weiler—now in charge of the Office of Space Science—would describe his problems: "I inherited a very sick program. Wes [Huntress] didn't know; it looked healthy when he left, but AXAF blew up in my face [when the launch had to be delayed a year], so that's my problem. You can't solve a $100–150 million problem by waving your wand around in the air."

We wrestled with our budget issues again on April 8 at what had become an annual management retreat, this one at the huge, drafty Stanley Hotel in Estes Park, Colorado. Each system agreed to establish an internal reserve of 5 percent to get us back to dealing with our problems internally rather than by appealing to the project reserves. The results were to be discussed in our June management meeting. At that meeting, responses varied widely—a few systems tersely reported that they had set aside the agreed money, with modest impacts. Others threatened to deliver less hardware or just said that they had not found a way to save. Simmons evidently felt that the exercise had raised everyone's consciousness to the cost issue and concluded the discussion: "It appears we have looked at our internal reserves and we are in fairly good shape. I propose we don't take any more action, but that we be conservative with our funds."

Meanwhile, JPL had become concerned about the load of projects that Charles Elachi, the director of Space and Earth Science Programs, was carrying and reorganized his department. An Astrophysics Flight Projects Office was created, under which we would fall along with a number of other missions. The director for this office was to be Larry Simmons.

JPL spent June and part of July searching for a new manager for SIRTF. The instrument principal investigators and facility and project scientists discussed the possibilities among themselves on June 4. The consensus was that we were entering the critical twelve months that would either get us to launch or bog us down in serious problems, and hence the selection was a very delicate issue. Tom Roellig summarized: "Larry has done an excellent job, banging heads to get people working together. There are some lousy managers. I'm concerned we might not get a good one. . . . The change could . . . result in loss of the teaming relationship and efforts by different elements to get to the manager first." Elachi promised to "make absolutely sure the person we select is one you are comfortable with." Three candidates were under consideration. Mike Werner, as project scientist, had interviewed the favored one and had very positive reactions. This candidate had extensive experience in software and integration and test, and had worked with Simmons on the HST instrument WFPC2. Simmons thought he would be a good choice as well: "He should be very good in the team environment. The other candidates are good managers, but wouldn't be so adept in maintaining the team culture."

The name of the new manager was soon revealed: Dave Gallagher. Because of a prior commitment to his family to take a vacation in August, however, Gallagher was not available immediately, and Simmons continued to run the project. He led the presentations at the quarterly review August 11 and ran the management meeting the following day. At these meetings we learned that IRAC had continued to fall off its schedule. The major issues remained as before: getting priority in the Goddard system for personnel to work on the instrument, interfacing with the JPL instrument control system, and getting deliveries from vendors. Tim Kelly also reported "storm clouds on the horizon" for the telescope. "We are holding schedule, but I have had to expend resources faster than I had planned to do it. It's like death from a thousand cuts. It isn't any one big thing, but there are lots of little problems that we have to solve, and they all take resources." IRAC was projecting the need for another $1 million, and Kelly thought his problem could be as big as $5 million but hoped to keep it smaller. IRS and MIPS were in the middle of a frantic hunt for replacement parts for their electronics, which had been found in May to be built on defective printed circuit boards. And Lockheed was continuing to lumber along on the spacecraft, progressing too slowly to make the scheduled mating with the telescope and demanding to be fed with large gulps of money from the reserve at regular intervals. Gallagher was going to have to make decisions fast.

13
THE FIRST HARDWARE IS DELIVERED

■ Simmons had set a high standard. He was able to motivate diverse groups to work together smoothly. His innovations included a wide-open management style, with key decisions vetted at monthly face-to-face projectwide meetings and open budget books. He had an effective personal style for dealing with controversy, often by stating, "I don't disagree with that." It was hard to argue back against this statement, although we all learned quickly that it was not quite equivalent to "I agree with that."

In addition to the reliance on teamwork in Simmons's management approach, the project elements were each using their own internal practices, and we were counting on a high level of individual responsibility to help keep costs down. The heavy reliance on the integrity of the organizations building SIRTF was not traditional for NASA. Normally for a large project like SIRTF, NASA inspectors would have monitored every step. Under the pressure of faster, better, cheaper, however, our approach was the only plausible way to hold costs down and speed up delivery of the hardware.

Gallagher had been hand-picked to continue in this style. Indeed, he appeared to operate in the Simmons culture more thoroughly than Simmons did. Simmons had occasionally shown flashes of irritation when confronted with unexpected problems. He had become particularly impatient with project elements that came to him requesting money from the reserve for problems they had created by poor planning or performance. Gallagher buried these irritations further below the surface. Simmons had tried to run the management team as a democracy, but only so long as he was pleased with its decisions. Gallagher spent more effort persuading the group to make the decisions he favored. Consequently, the project appeared to run almost placidly.

Gallagher continued and expanded another important Simmons trait: he maintained open communications and good relations with NASA Headquarters. Both managers were able to present our case and get help with problems, apparently without having to use hard-sell bargaining. Headquarters solved the problems they could as promptly as possible, and thus kept us on a relatively smooth trajectory toward launch.

Our rates of progress along that trajectory could be judged at our quarterly reviews. Each part of the project had drawn up a list of schedule milestones to measure its accomplishments. At these reviews we compared the number of milestones we had planned to achieve with those we had actually accomplished. For some parts of the project, the new style of operation was working extremely well. The IRS was officially completed right on time in March 2000, and after a modest adjustment was made the next month it was ready to be mounted in the dewar. MIPS came in soon after that.

As the leader of the MIPS team, I have far more insight into the development of that instrument than the others, although to the extent that Jim Houck and I have compared notes, the IRS approach seems to have been similar. I describe the process here.

Normally, a university and an aerospace contractor would form a team to write the original proposal for an instrument. With MIPS we had taken a different route: the scientists wrote the proposal with some assistance from a potential contractor (Martin Marietta) but with the understanding that we would have a competition to make the final selection. Our evaluation of proposals from Martin Marietta and Ball Aerospace ended up in a deadlock. To break it, team member Paul Richards, my deputy Erick Young, and I visited both. It was possible to see them in a single day because both were located in the Denver area. Ball filled a room with engineers who were knowledgeable about the technologies needed to build the instrument, and we had a lively discussion of the challenges we were going to face. Martin Marietta gave us a private room and ushered engineers in one at a time for us to interview. That was a mistake. Without the chance for their engineers to reinforce each other, shortcomings became painfully obvious. We knew by the end of the afternoon that Ball had won the competition. Paul rushed back home for another obligation. Erick and I drove to a small lake, strolled to its shore, and chatted about the situation as dusk fell. Just as we were about to turn back to our car, a perfect V formation of geese flew over our heads heading north toward Boulder and Ball.

The next ten years fell into a frustrating, repetitive pattern. We had to maintain a skeleton team at Ball. Every time we wanted to make a change or advance to the next step, and every time there was a major review, Ball would have to round up engineers who were engaged on other projects and convince them and their managers to help us out. Our continual starts and stops produced some cheerful cynicism in the on-again-off-again group. Our optical engineer used to say, "If it's May, we must be redesigning MIPS."

Under Erick's leadership we pressed ahead during this period with development of far infrared detector arrays to operate from 50 to 200 microns. They had to be a new invention; there were absolutely no other far infrared detector arrays in existence anywhere to guide us. The detectors had to work at about 1.5K, but amplifiers made as conventional silicon integrated circuits failed to operate to our standards if they were colder than about 20K. In the end, an amplifier was developed at Hughes Aircraft in Carlsbad, California, that operated correctly at the temperature needed by the detectors, allowing a relatively straightforward physical arrangement to connect detectors and amplifiers into the final arrays. Hughes Carlsbad was closed soon afterward, making our miniscule supply of the devices irreplaceable.

Detectors in general have a way of messing up the development of NASA astrophysics missions. University groups have had mixed success in providing detectors, but so have traditional suppliers. NASA operates in allegiance with the aerospace industry for nearly everything it does, though, and working in familiar ways provides a sense of security. When SIRTF moved to JPL, our role in the detector array development was questioned in many ways. Early on, JPL held a review of the program that apparently went smoothly. Behind the scenes, however, I received a letter directing me to find an aerospace contractor to build the detector arrays. Another time we made a critical advance in the detector array construction, but when I proudly described it at JPL, the project manager, without consulting any experts on the relevant technologies, forbade us to pursue the concept any further (an order I disregarded).

In the end, the pressure to reduce cost made the old JPL reservations moot. We could no longer afford to transfer our approaches to an industrial firm; we *had* to succeed at the university. We added engineers to our group with specialties similar to those the aerospace contractors would have provided. Simmons's project management philosophy allowed them to tailor their approaches to our situation rather than having to follow a prescription imposed by JPL. The ability to respond to the intent of the aerospace rules without having literally to obey the ones that were not applicable for our unique situation was key to getting the arrays built on a reasonable schedule. For example, since our arrays were to operate at 2 degrees above absolute zero, we simply waived rules about "burning in" the parts; there was no need to expose them to extended operation well above room temperature to weed out units with marginal tolerance of heat.

We also had to design, build, and test an instrument to receive the arrays. NASA puts a strong emphasis on documenting requirements and plans at the beginning of a project. We produced a series of small books, starting with the instrument requirements derived directly from our science goals and leading through a description of how we recommended the instrument be operated, to engineering requirements on the overall instrument, to the breakdown of those requirements into detailed requirements for the instrument subsystems (such as the detector arrays). All this writing forces a development team to think about the task from the beginning and to figure out the best way to do it.

Building something as intricate as a spaceflight instrument really tests good communications. In addition to all the writing, we had weekly telephone calls with the team at Ball and I visited them for a day or two every month. We made sure during the visits to have a lot of time for informal discussion; we took long lunches and even went on picnics. When nothing else was going on, I strolled up and down the hallways and dropped in on anyone available.

Figure 13.1 shows the basic design we adopted (a number of pieces have been omitted for clarity). The instrument is only about 40 × 30 centimeters. In common with many spaceflight instruments, it is sufficiently compact and convoluted that only an expert can see how the light actually gets through to the detectors.

We built up a modest budget reserve, which was critical to making steady progress. We spent some of it in ways not anticipated by the original plans. Our independence from the usual level of oversight by JPL encouraged us to anticipate problems and design solutions without seeking prior spending approval. In a more traditional management environment, any reallocation of our budget would have required lengthy negotiations, perhaps extending beyond the need date for contingency investments. In the end, we used a surprising number of these backup routes.

Everything was going unbelievably smoothly. We were far ahead of the project schedule with the electronics to control our detector arrays. In fact, we were so far along that Ball had supplied us with electronics built according to the flight designs to operate the arrays in our laboratory while we finished building the flight units. The IRS team was leading the construction of most of the electronics for both instruments, and they too were far ahead of schedule. Then, a detector controller electronics board stopped working in our laboratory. We sent it back to Ball. The engineers there deduced that there might be a problem with a "via"—a small, plated hole used to connect layers of copper conductor in the board. They cut the board open to look. Sure enough, the layers of the board looked like they had been subjected to excess heat and had come apart at the via. The board had been taken from the

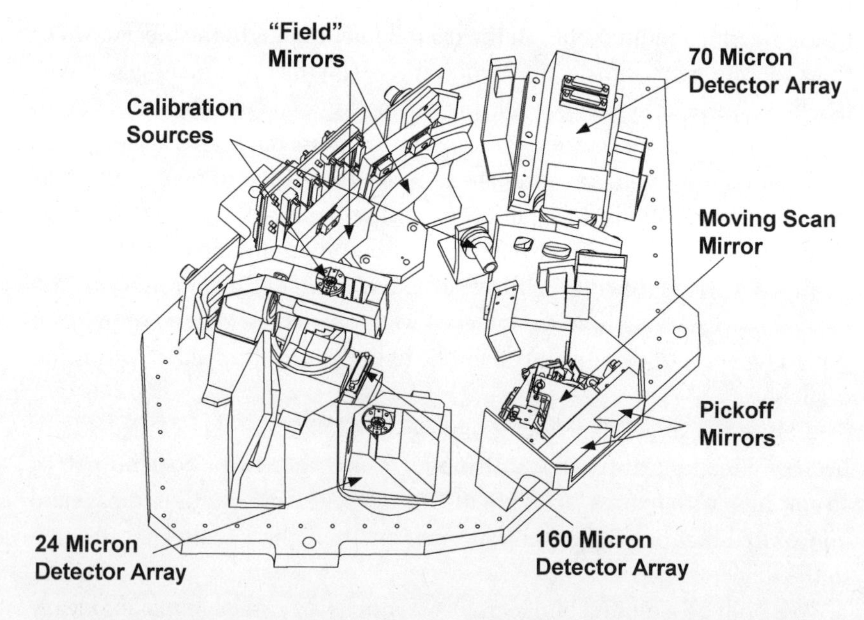

Figure 13.1. Design of the MIPS Instrument. In addition to the detector arrays, some key parts are: (1) the pickoff mirrors that receive the light from the telescope; (2) a scan mirror that allows the instrument field to be moved on the sky without repointing the telescope; (3) field mirrors that direct the light onto the scan mirror; and (4) calibration sources for the detectors.

same lot that had been used to build the ones we intended for flight. Ball sectioned other boards and found that the problem was universal, even though all had passed the screening tests at the manufacturer. We checked the second type of board we had built—same story. The IRS instrument team then looked at the boards for the rest of the electronics. All were defective. All the checks and tests used to certify the boards had somehow missed the problem.

The teams had just completed the flight electronics for both instruments, meaning that millions of dollars worth of hard-to-obtain electronics parts were now soldered onto our defective circuit boards. There was no choice but to throw everything out and start over. Both instrument teams began a frantic search for new space-qualified electronics parts and for board manufacturers who could make vias that did not have problems. We purchased boards from a number of suppliers and learned from the results that we had stumbled onto a previously unsuspected industrywide failing. Eventually, the IRS team found one acceptable supplier, and we discovered that changing the material for the electronics boards back to "primitive" fiberglass also resulted in acceptable products. With the basic problem resolved, the boards were replaced, but our nicely controlled development pace became far more intense for the final year of instrument construction and testing.

August and September 1999 were key to our success. By then we had to have the instrument fully assembled with the detectors and the optics aligned. The two teams, Ball on the instrument and Arizona on the detectors, raced to make the deadline. I spurred each team on with optimistic reports of the progress being made by the other. They ended up in a dead heat, both parts ready just in time to be mated smoothly and to continue the integration of the instrument. It was like a perfect football pass when the tight end catches the ball and turns toward the goal line without any perceptible change in stride.

The instrument was to be tested in a dewar Frank Low's company had built for us. We lashed together a set of electronics from the old, defective flight boards. Two young scientists had agreed to live in Boulder and take part in the instrument tests. Under the tutelage of the resident former professor and veteran of instrument testing, Bill Burmester, they quickly became adept.

Once the testing was completed there was a detailed audit in which we had either to show that the instrument met its requirements or to describe deviations from those requirements and submit them officially to the project.

Everything came together smoothly for the instrument's preship review, when we were to turn it over officially to JPL. Two days after the review, we learned that a detector array had failed and had to be replaced—fortunately we had a spare and were able to deliver the instrument in mid-May with no impact on the project schedule.

14 FINISHING THE DEWAR

■ Other parts of the project, such as IRAC and the spacecraft, were not achieving their quarterly review milestones on schedule. The overall situation with our spacecraft, and particularly with its software, was especially worrisome. Neither the previous reorganizations nor the substantial additional funding provided to Lockheed had managed to speed up the plodding pace. Lockheed abruptly dismissed both managers involved. Roger Grubic was eventually appointed to replace them. Grubic was forceful and blunt, with something of the demeanor of a military commander. Money was shipped to Lockheed in very large envelopes; eventually, the software team at Lockheed grew to nearly one hundred programmers, far larger than had ever been envisioned at the time of the June deal.

On August 9, 2000, we were at the quarterly review, held in the conference room under the watchful eyes of past JPL directors. Gallagher opened by promising to emphasize the two major problem areas: the spacecraft flight software and the late delivery of the IRAC instrument. Program Executive Lia LaPiana commented that the NASA Headquarters representatives were concerned about the possible delays that had been reported to them and "were very interested in establishing a realistic launch date."

Mike Werner made a brief report that the Legacy program opportunity, requesting proposals from the astronomy community for major sets of observations, had been released. Under normal conditions this news would have been featured because it would bring new scientific proposals into SIRTF for the first time since the instrument teams had been selected sixteen years previously. However, Mike's report was passed over quickly, as were the reports from two more project elements.

It was Grubic's turn for Lockheed. His news was surprisingly good. The new management team on the flight software was taking hold and had been through a successful review at the end of June. They now had finished defining the software requirements and had put together an integrated schedule with target dates for delivery of defined products. These steps are the foundation to managing any complex effort; they should have been in place a year and a half earlier and had been the top priority of the new team for the preceding six months. The board was pleased.

JPL associate director Tom Gavin congratulated Grubic on the progress and asked: "Are you still controlling the schedule?"

Grubic answered: "I think the IRAC activity will hold things up and we'll be ready by the time the project goes into thermal/vac."

Gallagher spoke up next: "I believe we've seen a significant turnaround in this area. Last quarterly we couldn't present an integrated schedule. Now we've had a peer review where the biggest comment was the schedule was aggressive, but Roger and I agreed we want to keep it aggressive. Lockheed has an internal process [when a program gets into difficulty] called 'return to green.' We're not there yet, but we're making progress. I'm very pleased."

The telescope presentation centered on good news about the image quality, but the board quickly turned to a discussion of the schedule and the impact of various options for dealing with IRAC, which was now the only remaining item standing in the way of finishing the assembly of the dewar and advancing toward launch.

It was time for the IRAC presentation. Giovanni Fazio made his way slowly to the front of the room. Fazio normally walks with a broad limp (and often uses a cane, but is sufficiently independent of it that he sometimes forgets it), but this time his progress was noticeably slower than usual. Eventually he stood at the viewgraph machine: "Now I know how it feels to be the star of a show. I'm wearing a bulletproof vest, so you can fire away." He tried to make the case for delivering only the cold part of the instrument that would be mounted inside the dewar: "As a result of the delays, we have acquired a lot of running time on the cold instrument, have been able to do long-term trending analysis. It looks very good."

Gallagher was skeptical. "This thing is so buried in the [dewar], you have to be sure you are ready to deliver—we keep finding problems."

Gavin was not convinced either: "If this were only a firmware problem, it would be low risk to deliver the hardware. But what I hear is there are unresolved problem failure reports that make you reluctant to ship. You need to concentrate on those issues, and I don't see this happening, guys. This firmware issue is a smokescreen."

The June deal had included more than budgets. Each project element had committed to delivering its share of SIRTF in a complete form at the time indicated on the schedule. That is, there should be no complications such as continuing to work on software while an instrument was being operated in the observatory. This agreement was a key part of Simmons's original phased development plan; only by holding each element to a strict schedule could we hope to contain costs. It had the further advantage that there would be a stable baseline for the integrated testing.

IRAC, however, seemed unable to commit to a schedule for delivery of the completed instrument: the statements by Dick Taylor (IRAC manager at the Smithsonian Institution) and Rich Barney (IRAC manager at Goddard) appeared to be contradictory, and Barney refused to name a delivery date. He complained: "We proposed to deliver with the scientific problem failure reports only [that is, to deliver the cold instrument separately], but that was not accepted. We do not have the other part [the electronics and firmware] or the verification of requirements from Lockheed [the check that the instrument meets its requirements]. There's a lot of history there. We have to resolve that and then we can finalize the schedule." Originally, a phased IRAC delivery had been rejected because of the possibility of finding problems with the cold instrument when it was run with the completed electronics and firmware. This set of tests was still far in the future because of delays in the firmware. Ironically, the cold instrument was complete, tested, and ready for delivery, but the incomplete firmware could be tested in a meaningful way only with the cold instrument; in effect, the cold instrument was needed as a test article for the firmware. The discussion of the firmware problems continued.

Dick Taylor: Testing with [the JPL-furnished computer system] and the approach Lockheed is taking with the verification have brought a lot of firmware issues to light. We have about forty of them now we have to resolve.

JPL *director Elachi*: Are you guys camped at Goddard?

Taylor: We've made some 163 trips to Goddard over the last nine months.

Simmons: What is the impact of warming up the instrument now? Test the firmware without the cold instrument.

Fazio: It would make me uncomfortable.

Simmons: So does delivery five months late.

Gavin: I understand. They can't ship until they're confident.

Barney: We only had one programmer, and he left. We had a plan to deliver the instrument with the four big firmware issues resolved, but the plan was changed [when JPL required a completely verified instrument].

LaPiana [to Taylor]: I hear Goddard speaking and you speaking, but I don't hear a single voice for the IRAC instrument.

Barney: Don't misunderstand. That's [pointing to Fazio] the single voice for IRAC, and when he speaks or Dave [Gallagher] speaks, that's what we try to do.

IRAC was finally set aside. There had been no resolution, just an open display of the difficulties in holding to the original rules for instrument delivery and exposure of some of the strains caused by Goddard leading the IRAC management and reporting directly to JPL while the principal investigator was at another institution.

There were short reports from a number of additional project elements, and then Gallagher stepped forward for a longer than usual overview of the project. He began: "We have a launch delay." Then he reported that the financial reserve was virtually gone—the progress in flight software had been expensive—$17 million—and starting from a $19 million reserve he nominally had only $2 million left with lots of outstanding demands on it. An intense discussion followed.

Ken Ledbetter [from NASA Headquarters]: We don't have any FY01 money. You have to stay within budget.

Gallagher: It can't be done.

Anne Kinney [the new head of the Origins program]: I am concerned that the IRAC effort doesn't look like a team. Charles [Elachi] asks, "What can we do to help?" That's a question you only ask once or twice, and then you step in. . . . The budget problems are going to be very hard for me to work. . . . SIRTF is often advertised as "the best managed" project. It's been my bragging point, and I need it to stay that way.

Gavin: It would be very bad form to announce a launch delay and then integrate IRAC and have to say, "Oh my, we've got a problem somewhere else." Go ahead and test the [dewar] now and fit IRAC in when it's ready.

The next morning the management team met in the conference room at the SIRTF Science Center on the Cal Tech campus. The upper three floors of an existing building were being remodeled to be used for the science operations of the project. The conference room had a pristine, unused look, and adjoining it was a large, similarly pristine room full of booths and small desks that was planned as the headquarters for testing the observatory when it was launched. The discussion began almost literally from where it had ended in the quarterly review, but now just among ourselves. Gallagher emphasized that "we only want to do this once." We would have to replan the entire mission for a delay, but we wanted the new plan to be conservative enough to cover all contingencies. We wanted it to be something that would stick.

When we had been struggling for our very existence and had a small model of the Delta rocket as our only launch possibility, we had adopted a plan for a "direct ascent" launch from the pad into space, with no second firing of the engine for trajectory adjustments. This strategy had saved us 80 kilograms of mass otherwise required for fuel to allow a second firing of the rocket engines. Although at the time most of us had been unaware of the constraints, we could reach the correct orbit from Florida with this type of launch only between about November 15 and February 15. Thus, we had almost no flexibility for launch slips unless we slipped a whole year.

Fortunately, we had eventually been given a much more capable form of Delta rocket, and so far we had committed only a fraction of its additional lifting capability. Thus, we now had some left—the rocket was capable of launching 940 kilograms, and we had committed only about 830 kilograms, so we could add the 80 kilograms back in and still have a thin margin left. The discipline to keep launch capability in reserve rather than to let mass creep up to the maximum possible had paid off. We could use a launch with a short coast stage—the engines would be turned off and the rocket and observatory would go into a "parking orbit" that would carry them partially around the earth. Once SIRTF was in an appropriate position, the engines could be fired again to put it into the correct orbit around the sun.

A vigorous discussion broke out about whether we could stay with the direct ascent plan. "November 15 is an eleven-month slip," Gallagher said, "and that ain't going to fly." Irace was convinced that we needed "to work the direct ascent [November 15 option] as a second window." Houck suggested that we "appoint a tiger team to see if there is a showstopper with the parking orbit." A "tiger team" is a small number of experts assigned to attack an issue intensely and quickly, an approach that circumvents the rather cumbersome way a large project usually moves to reach consensus. Gallagher agreed: "The action is to make a rapid assessment of the parking orbit. If it looks OK, go into the system design team and reconsider every assumption to be sure we haven't left anything out."

On August 24, the parking orbit tiger team reported that no issues had been found. Five days later, Gallagher kicked off our replanning for a later launch, hoping to "reestablish the risk profile of, oh, say, a year and a half ago." He emphasized that he wanted the plan to be a new foundation for getting SIRTF to orbit. He summarized the reasons to be more conservative in some areas: "From the day I joined SIRTF, I thought we had a pretty significant lien with only one cold cycle—all the reviews . . . said they had never seen a complex cryogenic system work on the first shot. We have an opportunity to increase the likelihood we won't have to go back and ask for more resources if we allow for an extra cooldown."

Gallagher would have to take our new plans to NASA Headquarters on September 28 to ask for approval and, along with it, extra money. An official launch date was selected: July 15, 2002. Each project element hastily put together a new budget to support this date. Gallagher was most interested in minimizing the extra budget needed in 2001 while at the same time keeping the project's "sense of urgency. The seven and a half months could be like a drop in the bucket. It has become clear we would have had a horrible time getting to the pad on the old schedule." Despite the protests from headquarters, it looked like we would need about $13 million additional in FY01.

On October 12, the management team assembled at Ball Aerospace and listened as Gallagher described the progress toward approval of our new plan. The pain in FY01 had been eased when it was realized that the launch delay had removed the need for spending on a rocket, allowing $11 million to be taken from the original FY01 budget and delaying some of the impact of the slip to later years. By using the money set aside to operate SIRTF in FY02 to complete it instead, the impact in that year had also been minimized. By this strategy, most of the problem was postponed to around 2007. If the launch was successful and the mission lived for a full five years, we would be short nearly $80 million in operating funds. It had been necessary for NASA Headquarters to take this sum out of other areas in their long-term plans. However, given the year-to-year pressures NASA works under, postponing a budget problem five years is a noteworthy accomplishment.

The previous day Gallagher had been back in Washington to meet with Ed Weiler (head of NASA Space Science), Kinney, and LaPiana. Weiler had approved the plans. Official notification would come later, but we were authorized to proceed on the new path. "I've never seen headquarters move this fast on anything," Gallagher said.

> We made the presentation on the twenty-eighth, and this is an $80 million issue for them. It's good news for SIRTF, but I've got to look at it from a program perspective, too. It will impact other programs, so I won't be bounding around JPL with the news.
>
> We have no reason not to succeed—we got the schedule we asked for, we got the budget we asked for, and we retired the liens. I wasn't around for the "June deal," but I want to emphasize this is a different situation. We have budgets prepared by experienced teams that have what they thought they needed. We've been given everything we asked for. We control our own destiny. I, for one, am committed to finishing the job as we have agreed.

Meanwhile, we had decided to go ahead and put IRAC into the dewar. We could not be sure that it was in good enough shape to remain there, but if we left it out, we were *guaranteeing* that we would have to take the risk of reopening the dewar.

In fact, the instruments were being mounted in the dewar only a hundred yards from where the management team was meeting. Late that afternoon a few of us went to look. We slipped into clean room garb—bunny suits—and entered the large central clean room. In one corner, under a separate protective tent, was the dewar, still open to our view. All three instruments were together for the first time. It was breathtaking. All the years of waiting, the painstaking efforts to make the instruments as perfect as possible, had culminated in the sight before us. And, at least to the trained eye, it was gorgeous. I silently said goodbye, knowing that the instruments would soon be sealed behind the dewar lid and hoping I would never see them again. Only if a problem emerged would the dewar be opened to gain access.

Weeks of checking and counterchecking passed before the dewar was actually sealed. For three weeks its contents were carefully "baked" to drive out water vapor. Then the air was pumped out and the dewar cooled with liquid helium. By early December it was apparent that it was working well. The middle two weeks of the month were consumed by rapid testing of all the instruments. Initially, one quadrant of an array in IRAC did not work, but the problem was traced to cabling external to the dewar and was fixed. The other two instruments and the pointing detectors all checked out fine, with only one possible exception. It appeared that the efficiency of the IRS was about 30 percent less than expected in one channel. But in general, it was becoming apparent that virtually nothing was wrong! We rushed to confirm this astounding result before Ball shut down for the holidays. Some last-minute trades in test time allowed us to complete the checks. Having just provided four months of schedule reserve against instrument problems, it seemed possible that we didn't need it!

On Thursday, December 21, we gathered over our conference telephones to confirm our good fortune. Each instrument was asked to describe its status. Everything appeared to be perfect, except for the minor reduction in transmission in the IRS. Almost reflexively, we descended into minor questions and picked at details rather than accepting the good news at face value. Realizing how precious it was to us now, Bill Irace worried about whether the dewar would be guarded over the holidays. After a bit of teasing from the rest of us, he dropped the subject to start wrapping up the call: "The mark of real quality is when a team makes something really hard look easy, and the Ball team has done that. I'm overwhelmed." Dave Gallagher joined in: "This is quite a way to start the holidays. On SIRTF we're pretty good at searching around and finding trouble, but we just can't seem to do that today." Our mission assurance manager, George Greanias, whose job was to find problems, added facetiously in the background: "It's really frustrating."

15

2000: THE RULES CHANGE

■ We had succeeded in completing the first major parts of SIRTF under our faster, better, cheaper–based management style. The IRS and MIPS teams had flourished in this culture. They had combined a university-based principal investigator team with an aerospace instrument contractor. The strengths of the university and industrial components were complementary in ways that allowed the combination to achieve the efficiencies that were the goal of the faster, better, cheaper approach.

The Mars Pathfinder project had demonstrated the potential efficiencies of faster, better, cheaper in 1997 when, on a small budget, it landed a rover on the planet and operated it there to the wonder of the entire world. The mission had combined successful science with a return in public interest as large as had ever been achieved per dollar invested by NASA. An accompanying mission, Mars Global Surveyor, was also highly successful, although less in the public eye. The Pathfinder success in the face of the faster, better, cheaper management challenges made it an icon for management innovation. Managers associated with the project teamed with professional writers to produce books and glossy publications that laid out the methods, with titles like *High Velocity Leadership: The Mars Pathfinder Approach to Faster,*

Better, Cheaper (Muirhead and Simon 1999). The breathless nature of one of these products is distilled on the back cover:

> Somewhere within, waiting for the call, there lies a creative spirit . . . a bit of the explorer . . . a love of challenge and personal adventure. Sometimes this part of our nature runs so deep and quiet we cannot feel its presence. Then sometimes it stirs, awakened by events that tap into this great potential. . . . These are among the high points of our lives. We truly live. We see more accurately how much more we can be and do. . . . The Mars Pathfinder team lived such an experience, and these pages reveal the secrets of how you can make it happen for yourself. (Pritchett and Muirhead 1998)

The full demonstration of the faster, better, cheaper approach had been due in 1999. The first test was to be the Wide Field Infrared Explorer (WIRE).[1] WIRE was of great interest to the SIRTF team because it was a cryogenic telescope carrying out surveys with detector arrays that had been the prototypes for some of those in the SIRTF instruments.

A prelaunch news release from JPL on February 23, 1999, set the stage. Titled "Wide-Field Infrared Explorer to Survey Starburst Galaxies," the statement noted that

> NASA's first new spacecraft in the Origins Program, the Wide-Field Infrared Explorer (WIRE), is scheduled for launch at 10 P.M. EST on March 1 from Vandenburg Air Force Base, CA. . . . "In many ways this inaugural mission of NASA's Origins Program, which will study the birth of star-forming galaxies, will move us towards our ultimate goals," said Dr. Harley Thronson, acting director of the Astronomical Search for Origins science theme at NASA Headquarters. "WIRE will provide us with a wealth of information, which will get us closer to understanding how the Universe could reach the point of forming Sun-like stars and Earth-like planets. And, WIRE will do that at a very modest cost."

However, on March 8 a news release reported disaster:

> Ground controllers are slowly gaining control of NASA's Wide-Field Infrared Explorer (WIRE), but the entire supply of frozen hydrogen

> needed to cool its primary scientific instrument has been released into space, ending the scientific mission of the spacecraft.
>
> "We are very disappointed at the loss of WIRE's science program," said Dr. Ed Weiler, NASA's Associate Administrator for Space Science at NASA Headquarters. . . .
>
> Spacecraft controllers believe the primary telescope cover was released about three days earlier than planned. As a result sunlight began to fall on the instrument's dewar, a container of frozen hydrogen designed to cool the instrument. The hydrogen then warmed up and vented into space at a much higher rate than it was designed to do, causing the spacecraft to spin. Controllers do not know what specifically caused the cover to be released.

More details were provided in June with the publication of the "WIRE Mishap Investigation Board Report" (1999). The electronics circuit designed to release the cover had been built with a flaw that caused an "open" command to be propagated briefly when the circuit was powered up. In this small interval the cover had blown off. Although the electronics designs had been reviewed, the faulty section had not been complete at that time and there had been no follow-on review—which might or might not have detected the problem.

Two faster, better, cheaper Mars missions—Mars Climate Orbiter (MCO) and Mars Polar Lander (MPL)—had been launched near the first of the year and were scheduled to arrive at the planet separated by about two months. On September 23, 1999, the ground control team at JPL decided against a final rocket burn to adjust the trajectory of MCO. Small anomalies had been encountered a number of times during the long coast of the spacecraft to Mars. As there was no worked-out procedure in place for a last-minute fix, however, it was decided to be too risky to try one. Five minutes after the beginning of orbit insertion, Mars occulted the spacecraft and radio contact was interrupted, never to be reestablished. The basic cause became apparent almost as soon as the first questions were asked. The spacecraft contractor, Lockheed Martin Aeronautics, had written the software in English units, while the controllers at JPL had assumed the units were metric and hence had planned all the orbital insertion maneuvers incorrectly. The spacecraft had probably spiraled down and crashed without ever achieving orbit.

There was a flurry of activity at JPL to identify any issues raised by the MCO failure that might threaten the success of MPL. A Mission Safety and Success Team that included more than fifty technical experts was formed. The team reviewed the critical sequence of steps required for MPL to land on Mars, including a detailed assessment of possible failures and a plan for circumventing each one. The team's report was released on December 1. In parallel, a "red team" was appointed to track all activities involved in the landing of MPL and to be sure all issues previously identified had been dealt with satisfactorily.

MPL approached Mars on December 3, apparently in good condition and ready for the complex series of events required to place its payloads on the Martian surface. The spacecraft turned to the attitude it would assume for entry. As an economy, it had been designed to operate from this attitude autonomously with no communications back to Earth until the lander was standing on Martian soil twenty-four minutes later. As expected, the spacecraft went out of communication, but nothing further was ever heard from it. The mission postmortem team eventually decided that a spurious signal had been generated when the spacecraft deployed its landing legs above the surface of Mars, suggesting to the control computer that the landing had occurred. The computer therefore shut off the rocket engines, and the spacecraft free-fell to a crash landing (Mars Program Independent Assessment Team Summary Report 2000).

Three prominent failures in rapid succession brought the entire faster, better, cheaper approach under question. To understand the issues, consider the challenges posed by science mission construction. It is amazing that the human brain, with capabilities honed through hunting antelope and avoiding tigers, can be so thoroughly reprogrammed as to design a scientific satellite. Although most satellites are successful, it is not surprising that bugs slip through. Weavers in Muslim countries are said to include an intentional flaw in every rug so as not to offend Allah by producing perfection. Allah will never be offended by a satellite. Despite all the care, despite all the checks and reams of paperwork, despite the seemingly endless reviews, it is simply not possible for the human brain to anticipate, identify, and correct all the possible design and construction problems in so complex a task.

Perfection is particularly elusive for science missions, which are built using a modern version of the medieval guild system. Aerospace companies identify individuals who are exceptionally talented in various design and construction tasks. These people make one-of-a-kind parts, largely by hand, to exquisitely high standards of workmanship. Since each science satellite is one of a kind, any issues that surface during construction must be dealt with ad hoc. In the end, everything is tested as thoroughly as possible, but a test program can probe only a limited set of conditions. The final success of the effort depends on the skill and integrity of the craftspeople who build the spacecraft as much as it does on the quality of the design and the test program.

In the early days of space exploration, the 1960s and 1970s, it was the custom to build an "engineering" model satellite to identify and correct design errors and to rehearse production techniques. The resulting improved design was used for both a "qualification" model and a "flight" model. The former was used for testing beyond the conditions that would be encountered in launch and flight, giving confidence that the flight model would work properly.

In principle, sophisticated computer-based design tools make engineering models unnecessary. The parts of a satellite can be based on accurate three-dimensional renderings within the computer that are transferred directly to machine tools that make them exactly as specified in the computer design. Similarly, the performance of electronic circuits can be simulated accurately before they are built. Computer analyses give greater confidence in the ability of a design to survive adverse environments without testing a qualification model. The easy way to stretch a budget is to build only one version of the satellite, counting on the computer-aided design to get the parts right and the further analyses to be accurate enough that the environmental testing can be conducted successfully on the flight article. A spacecraft built in this way is described as a "protoflight" model, a combination of "prototype" and "flight" that correctly captures the skipping of the intermediate models. Despite these apparent shortcuts, improvements in design capability and the quality of parts have led to a steady decrease in the failure rate in NASA satellites (Sarsfield 1998).

Prior to the changes instituted by Dan Goldin, increasingly elaborate analyses were conducted to identify potential failures. Reviews of ever-increasing scope and formality were imposed on the engineers and technicians putting each mission together. As the failure of Mars Observer and the problems with Hubble and Galileo show, however, even the most complex and expensive measures could not guarantee success.

That serious faults slipped by even those intricate regulations and reviews demonstrated a law of diminishing returns. The regulations capture the failures that have occurred in construction and operation of spacecraft in the past. Each spacecraft adds to the encyclopedia of what can go wrong. Yet, every new mission still exposes new kinds of risks. In general, each mission has only a limited reservoir of highly talented people who have the broad perspectives needed to catch and correct new kinds of potential failures. Their efforts are diluted if they have to spend time poring over an encyclopedia of old and possibly irrelevant entries.

The idea behind faster, better, cheaper was to produce high reliability in a different way. The checking and elaborate specification of procedures and approaches were reduced in favor of an increased reliance on the judgment of the project team. Elaborate analyses were discouraged if it appeared there was only marginal benefit. Reviews became less formal and more open, and were held on a schedule responsive to the perceived needs of the project. The historical emphasis on procedure was replaced with increased reliance on individual responsibility. Although these reforms are relatively simple in concept, they involved a large number of details in implementation.

The failures of WIRE, MCO, and MPL did not appear to result directly from the new approaches *intended* with faster, better, cheaper. Instead, they expanded on the historical trend for catastrophic failures in science missions to occur in the "simple stuff." That is, the transient behavior in standard control circuits for WIRE and MPL and the units confusion for MCO did not arise from highly complex aspects of the missions where detailed analysis and extensive, intrusive reviews might have exposed otherwise hidden weaknesses. They were first-order problems that should have been caught in even high-level reviews or testing.

It appeared that the guidelines for faster, better, cheaper were working for many of the "complex" aspects of the missions—areas that represented the newest technology and hence had the highest risk by traditional measures. The breakdowns were occurring as a result of the changed attitude toward the other areas. Thus, the Mars Program Independent Assessment Team (2000) reported: "Faster, Better, Cheaper (FBC) encourages taking prudent risk where justified by the return. [We] found that the lack of an established definition of FBC and policies/procedures to guide implementation resulted in project managers having different interpretations of what is prudent risk. Senior management needs to establish that risk associated with new high-return technology and innovation is acceptable as is risk associated with pursuing high-value science. Risk associated with deviating from sound principles is unacceptable."

Goldin admitted these shortcomings when he took responsibility for the problems. In an interview with Matthew Fordahl of the Associated Press on

March 30, 2000, Goldin admitted that he "pushed too hard, cut costs, and made it impossible for spacecraft managers to succeed." The interview continued:

> But Goldin said he will not abandon the National Aeronautics and Space Administration's "faster, better, cheaper" approach. Mission managers will get enough money and personnel to do the job, but there won't be a return to the days of big, expensive spacecraft.
>
> "We're going to make sure they have adequate resources, but we're not going to let the pendulum swing all the way back," Goldin told employees of NASA's Jet Propulsion Laboratory, where Mars Polar Lander and Mars Climate Orbiter were managed.
>
> Goldin visited the lab Wednesday, a day after two reports were released on the recent Mars fiasco. The reports found mismanagement, unrealistic expectations, and anemic funding were to blame as much as the other mistakes that doomed the missions.
>
> "In my effort to empower people, I pushed too hard," Goldin said. "And in doing so, stretched the system too thin. It wasn't intentional, and it wasn't malicious. I believed the vision, but it may have made some failure inevitable."

The general public probably has a more balanced view of NASA's successes and failures than can be held by a scientist involved in the program. The successes appear on the front pages and entertain and educate; their results become the subject of textbooks and part of our scientific legacy. The failures are sharply felt disappointments, sometimes making NASA a subject of ridicule, but they are eventually forgotten or eclipsed by another mission's success.

A scientist is in denial about possible failures when he writes a proposal and battles his way into involvement with a mission. Once he has invested many years of his career, however, the scientist becomes overly aware of the risks. Specifically, because of the long delays in getting SIRTF under way, we knew that few of us were going to have a second chance. The failures also hit NASA as an organization hard. There is an institutional level of denial that is most clearly revealed by the understated way failures are described, such as the WIRE "mishap."

The failures of 1999 sent the NASA management pendulum into a powerful swing toward conservatism. Abruptly, it was no longer fashionable to describe SIRTF as a faster, better, cheaper mission. Chandra, once the last of the dinosaurs because of its $2 billion price, was suddenly in vogue as a role model. After a year of delay to fix problems in its hardware and software, it had been launched successfully in July 1999 and was returning exciting new data.

JPL rearranged its management practices to provide more oversight of missions. We were transferred to a division under Tom Gavin, who as associate director of flight projects had the responsibility to oversee missions carefully as they approached completion. The next generation of missions from JPL was to be much more conservative in implementation and would cost substantially more as a result. Simmons summarized the change some time later: "We couldn't do SIRTF [today]. They wouldn't let us do what we did to streamline the project."

The increased caution created a dilemma for our project. As Space Science director Huntress had said after the MCO and MPL failures: "Better, faster, cheaper is here because we can't afford anything else" (Canizares 1999). Thus, our budget was not increased as a result of the new management attitude, although the cost cap was implicitly softened by statements from NASA Headquarters that we should not do anything stupid to save money.

But could SIRTF really be changed? Our initial approaches to defining the mission had in many ways been similar to those used in the two failed Mars missions. With high-grade hindsight, the failure reviews had found the Mars mission approaches to be aggressive almost to the point of irresponsibility. To a large extent these missions had been infected with their fatal problems in the design phases, such as the WIRE digital logic problem and the flaw in the software design and lack of a telemetry antenna for the landing attitude on MPL. Likewise for SIRTF, many of the basic decisions were already made and embodied in our hardware—or at least in the designs for it. We had no possibility of undoing much of the initial direction. At best, we could glue a thin veneer of greater caution and conservatism over the project; whether the veneer would only hide flaws or actually do some good would depend on how and where it was applied.

16

2000–2001: A CHRISTMAS PROBLEM

■ Following our triumph with the instruments and the dewar, Christmas 2000 was a wonderful holiday—a well-needed mental break from the stress of putting SIRTF together combined with the usual overindulgences in food, sleep, and merrymaking. Everyone tried to leave SIRTF behind at Ball Aerospace and concentrate on relaxation. Thus, when my pager went off on December 28, I was so mentally removed from the project that I did not recognize the number. When I punched it into the telephone, my manager at Ball, Scott Tennant, answered from his home. "There's a report that a burst disk has blown," he blurted out. I could feel myself grow pale. A burst disk is designed to break and relieve pressure in a dewar just below the point where the pressure level threatens an explosion. "Wh-where is the disk?" I stammered. "How much damage did it do?" "I don't know where it is, but they don't seem to be too concerned," Scott reassured me.

The next day we gathered around our telephones in a hastily assembled conference call. The failure had occurred on December 22 during a transfer of helium that was intended to put the dewar into a safe mode for the holidays. The engineer who had carried out the transfer had taken precautions to secure the system and had monitored it electronically from his home. Shortly afterward, he discovered that ice had frozen in the vent line and plugged it.

He had tried to free it, without success. The dewar was venting through an alternate path. He felt it was safe in this configuration and had notified no one else that there was a problem until the twenty-eighth.

Tim Kelly, as former telescope manager, had been called in to supervise the situation. He emphasized that he wanted us to concentrate on the immediate safety of the dewar, not on the longer-term problems: "I want to go through the issues. We have time-critical problems now. We should leave future reforms for the end or the next call. We are now 95 percent safe. We are nice and stable, but we have only one vent path, and I'm not going to feel safe until we've emptied out the cryostat."

Over the next two weeks we held a series of intense teleconferences. It was impossible to let go completely of our pre-Christmas euphoria. We clung to the possibility of clearing the ice plug without warming up the dewar so that we could proceed with our testing as if nothing had happened. The IRS team was particularly worried about warming up. They had decided that the reduced transmission in their instrument was due to deterioration in an optical filter, and their historical data indicated that it would probably get worse on every warm/cold cycle. The MIPS team had tracked down the cause of a failure of its flight spare 160-micron detector array. It too was related to the stress of cooling down. Although the flight array had been built in a way that reduced the risk, there was no way to guarantee that another warm/cold cycle was safe. The dewar now carried a working payload—after so much effort—and we had significant concerns about doing damage by warming it up. While our discussions droned on, the dewar sat in the Ball clean room, apparently venting peacefully.

On January 12, the newly selected Legacy science teams and the instrument teams met at the SIRTF Science Center. My pager went off in the middle of the meeting. I found a telephone and punched in the number. It was Scott again: "This morning, they evacuated the clean room and did an emergency vent of the cryostat." Scott had nothing more to report—he had been in the clean room working for another program and so his only involvement was as an evacuee. I went back and pulled Jim Houck out of the Legacy meeting. We called his lead engineer at Ball, John Troeltzsch. John had been working in the clean room doing tests, but he could add only the chilling rumor that the pressure in the dewar might have climbed to 75 pounds per square inch. The thermometers in the dewar had been indicating a problem since shortly after Christmas, but they had not been believed because the vent rate seemed to be about right. Finally, it had been decided that maybe the thermometers were right after all, that both lines had been plugged with ice, and that the only venting had been through the porous plug that is designed to let only a trickle of helium through on orbit. The "normal" venting was a combination of an abnormally large resistance to escaping gas and an abnormally high pressure to force the gas through the lines.

On January 18, a review was held at Ball to accept the SIRTF telescope. Once one of the biggest threats to the mission's success, now the telescope was a triumph: fully within specifications and ready to launch. Except for the cloud over the whole proceedings caused by the possibility that the dewar had been seriously damaged, the review went smoothly. That evening, Houck and I stayed late. Together with Bill Burmester, who was now in charge of instrument integration and test, we studied a crude model of the inside of the dewar that had been set up in the middle of the *Dilbert*-like labyrinth of engineer cubbyholes. We tried to understand what areas of the design were most vulnerable if the pressure had built to the point that the helium vessel had been permanently distorted. What had the pressure really been? Could it be estimated in a simple way from the physics of liquid helium? We went through a long passageway to the next building, where Mark Hanna, a Ball technician, was tending the dewar. Mark showed us the logbook where he and other technicians had conscientiously entered the readings from all the thermometers every hour, from the initial cooldown in December through the instrument testing and Christmas, and through the period when we were debating how to avoid warming it up while it was building pressure nearly to the bursting point. My stomach churned. The thermometers climbed steadily from the safe value of 4.2K through 5K, 6K, and above 7K. Mark was very upset—he had felt all along that the dewar was in trouble—but he tried to reassure us as we left. "Don't worry," he said, "we won't blow your instruments up."

The next day the management team met in the main Ball conference room. Gallagher wanted to move quickly and decisively on the dewar. He had decided to appoint a tiger team that would immediately start reviewing the data and would make an initial recommendation within a week on how to proceed. This recommendation would be submitted to the External Independent Readiness Review (EIRR) team. Only if both groups agreed would we proceed with our testing.

The four members of the tiger team spent most of the next week at Ball, and then reported back on January 26. They had used eight indirect methods to estimate the peak pressure reached in the tank and had decided that the highest likely value was 45 psi (pounds per square inch). Since the tank had been proof-tested to 53 psi at room temperature, and would be somewhat stronger at helium temperature, it appeared that we were in the clear! The team reported its results to the EIRR on February 1 and sold them on this conclusion.

On February 6 we were assembled for another quarterly review. Gallagher started it off:

> We are all focused on the cryostat [dewar]. However, we should not overlook the major, major advances that have occurred. The instruments are in the cryostat and they are all working. The telescope had a very successful acceptance review. The spacecraft is going well, and the star tracker had a good review.
>
> However, on the cryostat we had a major breakdown in our processes, and to correct them we are looking not just at the [telescope] delivery, but also at operations and processes at Lockheed and at the Cape. I've taken a series of steps to deal with the immediate problem. A tiger team . . . conducted a fact-finding meeting at Ball January 23 to 25. They reported to a formal failure review board February 1. There has been a reorganization of the Ball team. We will have a briefing to Weiler on the twenty-first and will verify the closure of the corrective actions on the twenty-sixth. Dick Hopkins will give you the details of what happened and what we plan to do about it.

Instead of Hopkins, however, Tim Kelly stood up: "Before Dick speaks, I want to say that Ball fully understands the seriousness of what happened. We will not only change the organization, but will do a lot of training not only in procedures but in how to work together." Hopkins then gave a technical summary of events. Vern Weyers, reporting for the EIRR/failure review board, followed him. Weyers endorsed the findings of the tiger team that we had not exceeded the proof pressure. Unfortunately, given how circumstantial the evidence was, neither Hopkins's nor Weyers's version was totally persuasive. Tom Gavin seized on a blip in the temperature profile near the end of the episode that had not been explained by the tiger team. He demanded that we find a solid explanation for everything before we proceed: "This thing has to be killed dead. You are not to change the configuration of the cryostat until you understand what happened. I agree that you might not find the evidence, but my Request for Action must be closed."

As the investigation continued, Gavin's hunch proved correct. Attempts to refine the calculations caused the pressure estimate to oscillate wildly, once climbing to 105 psi before settling in the mid-70s with a wildly optimistic uncertainty of 5 psi. We were stalled in deriving a credible pressure estimate. We were never going to know the pressure accurately enough to prove there had been no damage. We were still determined not to open the dewar; some other way had to be found.

Ball proposed to X-ray the dewar to see if there was any damage. We PIs made mother hens look totally irresponsible in the way we watched over our detector arrays. One small accident and our instruments would be crippled, and recovering would require months, maybe a year, and some good luck thrown in. We were all strongly opposed to X-raying our babies. Burmester worked hard to persuade us. He understood the risks better than we did and also appreciated our concerns. We trusted him because we knew that five layers of duct tape and a muzzle would have been inadequate to stop him from speaking out if he saw a problem. Under Bill's supervision, and with heavy shielding of the top end of the dewar where the instruments were perched, the X-rays were taken.

The management team met at Lockheed in Sunnyvale in March to review the dewar situation. Tim Kelly explained that the old estimate that we had reached 105 psi resulted from a faulty pressure measurement. Gallagher responded: "We're on the track of recertifying the cryostat in parallel to trying to track the pressure data, which is now a backup." Irace noted that we had to solve two problems: "(1) get the program going again; (2) convince every skeptic that we are successfully recertifying the cryostat."

The background to these positive statements was the X-rays themselves, copies of which were being handed around while the plans were developed. The remarkably clear images showed absolutely no distortions of the kind that would be expected from an overpressure. In parallel, a Ball mechanical engineer had painstakingly gone through each component of the dewar and computed the pressure it could withstand. This calculation indicated that the dewar could have withstood a pressure as high as 150 psi before damage occurred. With these two pieces of data we were going to appeal to standard engineering practice: if one did a proof test, then one had to inspect the vessel to make sure that it had not deformed. If the vessel passed the inspection, it was certified for use. We had just done a peculiar form of proof test. The X-rays constituted an unconventional but reasonably convincing form of inspection (some areas were cluttered due to overlapping parts). The test had required three months for analysis and discussion, thus consuming the schedule reserve we had set aside for more conventional problems in the dewar. It was just as well that the instruments had all worked; if not, we would already have fallen far off our new schedule.

An important aspect of the dewar incident (and of a number of other problems we encountered) is how Gallagher focused the project on resolving the problem and moving on, not on finding and punishing the guilty. In fact, although there were some obvious scapegoats, the guilt was widely spread. We had all listened to the claims that the temperature readings were untrustworthy and had not questioned them. We had all been too ready to put the dewar in a corner and give it minimal attention during the holidays. Bill Irace had been closer to the mark than any of the rest of us when he had pointed out how valuable the dewar was and had questioned whether we should post guards.

By pressing ahead we avoided defensiveness that would have hindered recovering from the incident. The concentration on recertifying the hardware minimized the impact on our schedule. Ball added a number of high-level people to the telescope team who helped to reform the procedures for operating the dewar and defined steps to deal with any future emergency. Once the procedures were in place, we could say that this particular crisis was behind us. The real question was how to avoid the next potential accident.

In April 2001 we were back at the Stanley Hotel for another management retreat. We had the usual discussions about the progress of the project, spent a little time hiking together, had a wine tasting, and watched *The Shining* (a horror movie based on a book written and placed at this hotel). In addition, two men grandiosely styling themselves "The Error Prevention Institute" (EPI) instructed us in the causes and possible prevention of mistakes like the one we had made with the dewar. They characterized one type of weakness—"the power of an idea"—and showed a re-creation of a real incident involving the crew of an airliner who were fixated on an altimeter that they were convinced was reading wrong because it contradicted the autopilot. As they discussed it, tapped on it, tried to bring it back to life in any way they could, it inexorably wound down toward zero altitude. The confusion came to a violent end when the airliner crashed into the Everglades—the altimeter had been working all the time. Fixated on an incorrect idea, the crew had ignored the malfunction of the autopilot. This incident recalled sharply our weeks of fixation that the dewar temperature sensors could not be trusted, leading to the conclusion that the helium was not at the elevated temperature the sensors indicated and that everything was venting perfectly correctly. Luckily, we had pulled up just before a violent crash.

The EPI instructors told us that after almost every disaster it turns out that someone in the organization had had a premonition that something was wrong, just as had occurred in our case. The problem was that communications in most organizations are hierarchical, and this premonition is seldom shared—they cited studies showing that in the average organization there is one useful suggestion per employee per seven years. We should have held what they called a "huddle" to bring together everyone with possible insights and then encouraged them to share their concerns, even ones that they could not fully justify, termed a "pinch," similar to Irace's request for guards.

There were other near misses as the project fell further and further behind schedule. The enormous pressures to earn a high percentage of the incentive fee, to avoid being the scapegoat for a launch slip, and to keep off the critical path (i.e., being responsible for schedule slips of the overall project) in a large project such as SIRTF can override common sense.

17

2000: SELECTING SOME SCIENCE

■ As the instruments and dewar were being put together, we began selecting the science projects that would actually use them at the beginning of the mission. In fact, our instruments were formally viewed as the means to carry out science investigations. Our ability to design, build, and deliver an instrument had been one consideration when the feasibility of our proposal had been evaluated back in 1983; our qualifications to carry out the research proposed had been another. Other members of the Science Working Group who were not responsible for delivering a major part of the mission hardware had written proposals even more strongly emphasizing the science. As payment for two decades of headaches we had been awarded observing time for our own projects. The exact amount had been renegotiated a number of times as the mission design changed, but it had finally settled on 5 percent of the total time per instrument team. Another 5 percent was distributed among the other Science Working Group members. The science center director retained another 5 percent, and the remaining 75 percent was to be allocated by competition among the astronomy community.

In acknowledgment of our long-term personal investment in the mission, the Science Working Group members were allowed to pick their science targets first, and to reserve them so that other astronomers could not collect similar data. The measurements would be kept "proprietary" for a year after they had been obtained, giving us a chance to reduce them, understand them, and publish papers on them.

We had gotten used to negotiations over money; relatively speaking, those discussions were routine because the negotiables were tangible. The project was buying our services for managing the instrument construction, supporting reviews, or developing software, and the quality of our performance depended directly on the resources they could provide. When it came to the science, however, we had to work out arrangements among ourselves based on intangibles. "Intangible" because no clear monetary value can be placed on a science investigation; any two scientists who tried to assign one would almost certainly come up with different numbers. "Intangible" too because the rewards we were hoping to reap for our years of work were abstract: to make some lasting contribution to human understanding. In addition, because of the huge advances in capability with SIRTF, the boundaries of an investigation were not clear. While one promising science program might wither unexpectedly, another might blunder onto a discovery and blossom far beyond anything originally envisioned.

The easiest way to deal with these intangibles would have been for each group with a small piece of "guaranteed time" to hoard it, not revealing their plans to anyone else, and then surprise the world with finished reports on any discoveries. Proceeding in this way would also diminish the returns, however, because the project would not benefit from the expertise of other instrument groups and scientists. We had therefore gone through an awkward and protracted process to establish collaborations, mixing negotiations with workshops and outbursts of annoyance, and awkward periods of silence with dispassionate discussion. A pattern gradually emerged. Our initial approaches to coordinating science programs were defensive, and we thought of almost legalistic solutions. There might be a demand such as: "Let's work together, but a member of my team has to be the first author of the first five major publications, and you will not be allowed to see our original data; send us yours and we will combine the two sets." The other team's reaction to such team-centric proposals was usually annoyance followed by a period when nothing seemed to happen. But fundamentally, both sides could see substantial advantages to working together, so someone would think of a counterproposal: "You can have first authorship of the first paper if we get it on the next ten; and let's appoint a subgroup of each team that will work to combine the data." Once the teams had groped their way to some solution to the legalisms, members really interested in the particular scientific problem were permitted to work together unsupervised by the rest. The painfully negotiated legalities would quickly be set aside (although they were presumably still in the background) and a carefully optimized program would emerge, bridging instrument boundaries and often bringing in key experts from outside SIRTF. This pattern became so predictable that it was tempting to sign up for anything in the legalistic stage, because the conditions would almost certainly be soon forgotten. The set of programs that emerged from these negotiations included a huge range of astronomy and drew nearly every member of the Science Working Group into a web of collaboration that included the four defining areas we had used to set the design of the mission but also many other studies that could use the same capabilities.

The Legacy science program gave members of the astronomy community their first opportunity to obtain observing time. It had been invented in reaction to the biggest shortcoming of the SIRTF concept: its limited life set by its finite stock of liquid helium. Once the helium had boiled away, the observatory would become a piece of space debris, its detectors too warm to operate correctly.[1] Because SIRTF would be so much more powerful than ground-based telescopes, once this happened we would no longer be able to detect, much less study, many of the sources it had found. Further, under our original plan with the one-year proprietary period, there would be virtually no chance for a second look at any SIRTF discoveries within the nominal two-and-a-half-year mission. Most of the scientific results would emerge only at the end of the proprietary period. By the time new proposals based on the discoveries could be written, reviewed, and accepted, and the commands to the observatory generated, SIRTF's helium would possibly be gone. Although many of these delays were fundamental, the proprietary period was not. A plan had been developed to accept proposals for "large and coherent science projects, not reproducible by any reasonable number or combination of smaller . . . investigations . . . of general and lasting importance to the broad astronomical community with SIRTF observational data yielding a substantial and coherent database." In return for early access to SIRTF data, the teams proposing these projects had to agree that their data would "enter the public domain immediately upon SSC processing and validation, therefore enabling timely and effective opportunities for follow-on observations and for archival research, with both SIRTF and other observatories" (Spitzer Science Center 2004).

In November 2000, six projects were selected from the twenty-five or so Legacy proposals submitted. In all, they would use nearly half a year of observing time. It was a relief to see such strong community interest in our mission. Some of us had feared that the long series of descopes had so reduced SIRTF's versatility that only a few applications beyond those represented in our four defining themes would be offered. The projects submitted by the guaranteed-time observers (GTO) and Legacy programs and the Legacy proposals that had been passed over were already enough to keep SIRTF busy for the entire two-and-a-half-year nominal mission. In addition to the four defining science programs we had originally planned, two Legacy programs addressed other issues: infrared properties of our Milky Way galaxy, led by Ed Churchwell of the University of Wisconsin; and infrared properties of nearby external galaxies, led by Robert Kennicutt Jr. of the University of Arizona. However, below we concentrate on the four defining programs.

In the seven years since we had adopted the brown dwarf theme, the status of these objects had been transformed from rare to commonplace. Astronomers were now engaged in classifying various types of brown dwarfs rather than debating their existence. It was also obvious that they formed in reasonably large numbers, and all the star-formation theories that had predicted that there would be few of them had been abandoned. Many theories now predicted that the stellar-based formation process might fail for objects less than about 1.5 percent the mass of the sun, or fifteen times the mass of Jupiter. The ISO observations of brown dwarfs had largely been outstripped by measurements from the ground. However, the SIRTF detector arrays at the critical wavelengths near 5 microns were both two orders of magnitude larger and two orders of magnitude more sensitive than those on ISO. SIRTF GTO and Legacy programs accordingly planned to search for young brown dwarfs down to five times the mass of Jupiter or less. By learning how many of these very low mass objects had formed, we could probe the new theories of how interstellar clouds fragment into stars.

In addition, IRS spectra could probe the nature of known higher-mass brown dwarfs. Ground-based studies were doing very well in the spectral range out to about 2.5 microns. However, many important spectral features lie at longer wavelengths, where ground-based instruments are blinded by the emission of the telescope. IRS observations would measure the behavior of the fundamental bands of the hydrocarbons that dominate much of the behavior of massive planet atmospheres—the "weather" on these objects. At even longer wavelengths, where the molecules in the brown dwarf atmospheres have few spectral bands, we were going to take observations with MIPS to measure their true temperatures. Determining the rate at which their trapped energy was escaping would help us determine their ages and masses. Together these studies would advance our understanding of the masses, ages, and atmospheric behavior in the brown dwarf types already known.

A great deal of time would be dedicated to debris disks, another of our four defining programs. Various samples of nearby stars would be surveyed at 24 and 70 microns to search for debris to much fainter levels (by a factor of about one hundred) than were reached by IRAS and ISO. The studies were organized loosely to use the new data to probe the properties of the debris systems as functions of (1) the stellar age, (2) the presence of companion stars or massive planets (the latter judged from ground-based measurements of subtle velocity shifts as the planets orbit their stars and tug them back and forth with their gravitational fields), and (3) the mass of the central star. Two Legacy teams had been selected for large programs. One, under Neal Evans of the University of Texas, would concentrate on regions where stars are currently forming to "follow the evolution of disks from starless cloud cores to planet-forming disks" (Spitzer Science Center 2004). That is, they planned to map large areas containing dense molecular clouds in order to find molecular cloud cores just starting their collapse into stars. These maps would also capture young stars with disks where many small planets had probably condensed out of the gas and dust and were beginning the process of combining into larger ones through multiple violent collisions. These stages and the ones in between are thought to occupy roughly the first three million years of the life of a star. The second team, under Michael Meyer of the University of Arizona, was to start with objects of this age and then

> trace the evolution of planetary systems at ages ranging from: (1) 3–10 million years when stellar accretion from the disk terminates; to (2) 10–100 million years when planets achieve their final masses via coalescence of solids and accretion of remnant molecular gas; to (3) 100–1000 million years when the final architecture of planetary systems takes form and frequent collisions between planetesimals produce copious quantities of dust; and finally to (4) mature systems of age comparable to the sun [5 billion years] in which planet-driven activity of planetesimals continues to generate detectable dust. (Spitzer Science Center 2004)

The ISO spectra showing crystalline minerals in a few circumstellar disks suggested a new direction for SIRTF. The IRS was scheduled to obtain spectra of many debris systems to see if these mineral features were common. The new data would help us understand the violent early stages of the planet-forming process, when planets first start to collect out of the material surrounding a young star, when collisions between young planets must be common, and when crystalline materials may be created in the searing heat of these events.

The IRS team planned to take spectra of ultraluminous infrared galaxies, our third defining program, at much higher sensitivity than had been possible with ISO. They would detect the emission lines in these objects that could provide unambiguous tests of whether they were powered by active galactic nuclei (AGNs) or by star formation. They would be able to explore this behavior to large enough redshifts to see if the sources were systematically different at an earlier stage in the development of the Universe.

X-ray astronomers had brought us another AGN-related puzzle. By comparing the individual sources they could detect with the diffuse X-ray emission of the sky not connected with recognizable objects, they had concluded that most of the X-ray emission in the early Universe is due to young AGNs. In most directions, the outputs of these objects are very strongly absorbed in the optical through soft-X-ray regions by gas and dust around the galaxy nucleus, making them nearly invisible to most X-ray surveys. It is unsatisfying to have theories constructed on the basis of sources we cannot see. However, in these sources, the absorbed energy would heat the gas and dust, making them bright in the infrared; also, the redshifted light from stars in the AGN galaxies would appear in our bands. If so, our deep surveys might find a population of relatively bright infrared sources associated with very faint X-ray ones.

Progress since our "Broomfield Accords" suggested a number of breakthroughs we might make on the fourth program, defining the behavior of galaxies in the early Universe. Optical astronomers using deep Hubble Telescope surveys claimed to have mapped out the pattern of early star formation, which they viewed as being in the form of bits and fragments of ultraviolet-bright star-forming objects that had not yet merged into large galaxies. Submillimeter astronomers had come up with a very different view. They had found a population of distant galaxies with huge energy outputs emitted in the far infrared (and redshifted to the submillimeter). These objects appear to be of high mass, perhaps even in the act of merging to grow further. They seem to contain large amounts of dust that absorb the visible and ultraviolet photons and emit their energy in the far infrared, making them nearly invisible to Hubble and accounting for their large far infrared fluxes. With a factor of one thousand separating the wavelengths for these two sets of data, it is not at all surprising that they see the early Universe differently. SIRTF would neatly fill this gap to bridge from the optical to the millimeter wavelengths, with the potential to unify the disparate views.

ISO had made deep maps to study sources in the early Universe at 15 and 175 microns. These data pushed the study of infrared galaxies out to high redshift and indicated that the picture of star formation derived from optical data was incomplete. However, because the ISO detector arrays were very small, only relatively small regions had been studied to deep limits; in addition, the spectral region between these two wavelengths was probed only to levels moderately deeper than had been reached by IRAS. To improve this situation, a combination of programs was planned to provide a set of surveys of a large area of "blank" sky. The surveys would have a progressively narrower focus. Starting with a large-area survey as a base, each subsequent step would examine a smaller area, probing it more deeply, until the fundamental depth limit of the observatory was reached. A Legacy team led by Carol Lonsdale of the Infrared Processing and Analysis Center at Cal Tech would survey the large area, roughly 50 square degrees, or three hundred times the amount of sky covered by the full moon. By necessity, this large-area survey would be relatively shallow, with the aim to identify large numbers of relatively bright sources. The team hoped to "trace the evolution of dust, star-forming galaxies, evolved stellar populations, and active galactic nuclei, as a function of environment from $z = \sim 2.5$ to the current epoch . . . [and] address the clustering of evolved stellar systems versus active star-forming systems and active galactic nuclei in the same volume" (Spitzer Science Center 2004). The guaranteed-time observers would examine the next segment, starting with a field observed to a depth similar to the Lonsdale program but only about one-sixth as big, and ending with an area a couple of square degrees—a few times the area covered by the full moon—surveyed three to four times deeper. A second Legacy team, led by Mark Dickinson of the National Optical Astronomy Observatory, would concentrate on very small area but very deep surveys. They hoped "to detect the rest-frame near-infrared light from the progenitors of galaxies like the Milky Way out to $z = 4$, and . . . to measure the stellar mass distribution of galaxies through most of cosmic history" (Spitzer Science Center 2004). The fields to be studied were selected because they already had good coverage with HST in the optical and with

Chandra in the X-ray. The SIRTF sources could be related directly to what was being seen in other wavelength regions, as had originally been envisioned in the Great Observatory concept. The full set of surveys, and IRS spectra of sources found in them, would characterize the infrared-emitting sources in the early Universe to the limits achievable by SIRTF.

18 NEW PROBLEMS

■ On the evening of May 8, 2001, I flew to Los Angeles for the quarterly review and management team meeting. The flight got in early, and I was provided with a higher grade of rental car than usual. I drove to Pasadena and went to Miyako's for dinner. Miyako's is a classic and venerable Japanese restaurant for whose shrimp tempura I had acquired an addiction. The tempura was as good as ever, and I indulged in a large bottle of Sapporo as well. I had become superstitious about fortune cookies. At the end of the meal, I broke mine open carefully so as not to damage the scrap of paper. It read: "Your impossible dream will soon become a reality." We had survived the dewar mishap and were proceeding smoothly toward launch. Life was good.

Early the next morning we all assembled for the quarterly review. The instruments had been moved to the top of the agenda, and IRS to the top of the instruments. Most of the IRS presentation focused on the failing filter discovered in the first dewar test. Jim Houck explained how the 30 percent transmission reduction in itself was not an insurmountable science loss. However, the manufacturer had now determined that the problem was a separation of the first interference layer from the substrate, meaning that a thin stack of filter layers was dangling without proper support. Houck was now concerned about the possibility of a catastrophic failure of this stack, which would produce many tiny pieces of the filter layers.

Nick Vadlamudi, the head Lockheed systems engineer, previewed another impending and still underestimated problem. Particles were contaminating the gas jet valves we relied on to help maintain the spacecraft's attitude (the "reaction control system"). Lockheed was adding filters to the gas lines to catch the particles.

Roger Grubic reported next: "Except for the reaction control system filters Nick just mentioned, the last piece of flight hardware delivers this Friday." It was the end of an era. Construction was complete, and we were purely into integration. "We are very close," Grubic continued, "only months away from having flight-ready software. We have completed builds 4.0 and 4.1; we still have to complete the fault protection code and the final review. Given what we have here [pointing to his projected chart], we have eight to nine months to run the software and review it."

Grubic's report was greeted with skepticism. Program Executive LaPiana spoke first: "You've slipped six months in the last eleven—how sure are you that this plan will hold?" Grubic responded: "I'm absolutely confident that we will have fault protection in place by the time the [telescope] is delivered in October." Gallagher offered support: "We had a formal review of the flight software plan eleven months ago. The opinion of the review board was that the plan was overly aggressive and we wouldn't make the dates. Roger and I discussed the plan and decided to hold to the aggressive timeline. It has turned out we did the right thing, and the slip did not just come out of the blue." LaPiana was not convinced: "I'm very concerned about the six months. One or two would have been OK."

In the review board's summary, the IRS filter topped the list of concerns. Harley Thronson (an infrared astronomer now at NASA Headquarters) described the situation: "We're about to launch a Great Observatory with a known flaw in its scientific capabilities. NASA has never done that before, and the implications are profound." Ann Kinney (head of Origins) struck a different note: "We are entering the most stressful part of the project. My concern on the IRS filter is the same as yours, Dave—have we found the last problem? We've already spent two silver bullets, one for IRAC and one for the cryostat, and they're getting very precious."

Gallagher had asked a small group of us to have dinner with him that evening at Café Bizou, a seafood restaurant in Old Pasadena. "We" included the three instrument PIs, Tom Soifer, Mike Werner, Charles Lawrence (Werner's deputy), and Bill Irace. Larry Simmons joined us a bit later. It was a pleasant evening, but it was not a social event. Houck had prepared placemat-sized lists of the status of all the issues related to the broken filter. Gallagher introduced the discussion, and Houck then started to take us through the items on the placemats. Somewhere in the middle, we ordered dinner. The food was delicious, but no one commented on it. We spent the whole dinner, and lingered afterward, trying to decide on a course of action. It was a gut-wrenching dilemma. Should we stop now, open the dewar, and replace the filter? Should we stay the course? Should we go through the vibration test needed to qualify the dewar for launch and then decide? Opinions varied; even ways of approaching the problem varied. Gallagher thought of it in terms of the risk to the mission if the filter disastrously disintegrated. If the filter continued to get worse, would we end up with a cloud of small particles (Houck estimated a million) that would eventually leak out of the IRS module and spread over the instrument chamber, scattering and blocking light for all the instruments? If that was the likely result, Gallagher felt we would have no choice but to open up. However, if the filter was "safe" in terms of the rest of the mission, he was worried about the consequences of stopping for the four to five months required to change it. There were other issues: would we damage other instruments by opening the instrument chamber? If we relaxed the schedule so soon after the dewar-induced delay, how would we make it to launch? Were there other problems lurking that would emerge later and cause further delays?

In contrast, Bill Irace tended to emphasize the scientific implications, subtly challenging the scientists to demand that the filter be changed for the good of the mission. Although Houck and Charles Lawrence were ready to rise to this challenge, none of the rest of the scientists felt so strongly. We were torn, on the one hand, between the desire to fix the filter, get IRS back to full performance, and eliminate the mission risk, and, on the other hand, the possibility of damaging existing hardware and the political dangers of such a large delay and added cost. At one point, Mike Werner described a report he had commissioned on contamination in an instrument on the Hubble Telescope. Irace dismissed the results, provoking Mike into a very uncharacteristic annoyed response. We searched for any certainty but found none. Gallagher took a few notes on a crumpled piece of paper. Eventually, Houck drew a scorecard on a piece of paper and went around the table asking for a vote from each of us. He placed a large dot in the appropriate column of the scorecard, except for a few particularly inconclusive opinions that he annotated with a small drawing of a waffle. The picture was clear at the end. The consensus was to proceed through vibrating the dewar, and if no changes were apparent in the filter transmission after vibration, to proceed through launch in the hope that nothing disastrous would ensue. Only Houck and Lawrence favored replacing the filter immediately. Simmons did not vote, arguing that it was our project, and his job was to help us implement any decision we reached. Irace voted either to change immediately or never to change at all and was rewarded with the largest waffle.

The next morning the larger management group—twenty-five to thirty people—met in the JPL project "design room" to continue the discussion. Gallagher read his list of critical issues from the crumpled sheet of paper:

1. Attempt to quantify the risk of catastrophic failure—if it is a significant safety issue, everything else is moot
2. Assess the likelihood of further degradation
3. Have a clear understanding and articulate well the science impact
4. Have a clear definition of the technical plan, cost, and schedule—be prepared to prove a change of filters could be done safely
5. Need a plan to present to headquarters that we can defend, no matter which way we decide to go

Houck gave us a detailed description of the filter, its mounting, and how it had been tested to qualify it to be mounted in the instrument. He felt the separated filter layers were almost literally hanging by a thread, and he emphasized the high degree of stress on them because of the differing properties of the layer materials. Nick Vadlamudi observed: "I don't have confidence if a part is off nominal that going through vibe and surviving will mean it is safe for the next vibe. Using the vibe as a gate is not the right approach." In other words, he questioned whether any conceivable test could "prove" that the filter would not fail later. Houck argued: "If we decide we won't take the schedule hit to change the filter, and in a few months we have to swallow a four-month delay, I'm going to have to take a lot of tranquilizers." Gallagher worried about the delay that would result from changing the filter when otherwise "it is perfectly feasible that we make a July launch. We have a lot of problems, but also a two-month reserve, and everyone still thinks they can make the schedule—Roger, Bob, and Tim, they all agree."

We decided to go ahead with the vibration tests, to make sure that any other problems had been found. Then we would take another look at the filter transmission and decide what to recommend. We might find it anywhere between unchanged and completely destroyed, but it was clear that no matter what its condition we would still have to argue our way onto a path to pursue. By our decision, that discussion was postponed for another month, during which we might hope to add a few more facts to our small supply of them.

As we progressed through testing, we discovered that the IRS filter was only the first problem affecting the instruments. None of the issues caused total loss of any scientific function. Following the precedent set by the IRS filter, as a group we were willing to accept the compromises in performance resulting from the various problems, although each one was painful to the affected party. Thronson's concern that we were "about to launch a Great Observatory with a known flaw in its scientific capabilities" proved optimistic: we were going to launch with a number of such flaws.

IRAC included a shutter to help with calibration. As one step toward finishing its control firmware, the order of operations for the shutter was changed —and the shutter then stuck in the closed position that blocked the instrument from viewing the sky. To keep the power requirements low (since any power dissipated in the shutter boiled off helium and shortened the mission life), the shutter had been built with a delicate balance between electrical torque to close it and a spring to open it. Evidently, some small extra force had developed that held the shutter closed against the force of the spring. It was found that the tendency to stick depended on which set of electronics was used to operate it. Working from this clue, the issue was duplicated with another unit in a laboratory at Goddard. It appeared that the electric current used to close the shutter also magnetized the shutter structure slightly. When the structure was assembled to normal tolerances, the contact area between fixed and moving parts was small and the magnetic force was not important; thus, no problem had been detected in the development unit. However, an extra-careful alignment had been carried out on the flight unit, which increased the contact area enough to allow sticking. The situation was one of the clearest illustrations of the wisdom of the common saying in spaceflight work that "there is no such thing as a small change."

The IRAC team investigated the problem and developed a strategy by which, they were convinced, the shutter could be opened under any foreseeable circumstance. If the shutter should stick in flight, however, it would be impossible to acquire any data with the instrument. Gallagher felt that they could meet their calibration requirements without the shutter, and therefore that any use was an unnecessary risk. He directed that it not be used. Additional data on the calibration strategy without the shutter supported his view.

The next casualty was MIPS. The signals from half of its 70-micron array disappeared during testing for electromagnetic interference, and then just as mysteriously reappeared a few days later. The failure recurred when we tested in an environmental chamber at Ball, and heroic efforts finally uncovered a faulty joint in a connector in the cabling between the cold instrument and its electronics. Unfortunately, during the environmental chamber run another failure appeared and then disappeared in one output (of eight) of the 160-micron array. From its nature this fault also had to lie in the cabling between the instrument and its electronics.

So great was the fear of injuring the carefully assembled telescope and dewar that there was considerable debate about whether to attempt to fix these cables, even though they were likely to lie outside the dewar and could be reached after only a couple of days of work. Eventually, Gallagher directed Ball to disassemble the telescope partially and to replace the accessible portions of the two cables in question. This part of the cabling included the connector known to cause the problem at 70 microns. The removed cable portion for the 160-micron array was only a part of the total cable run to the instrument, and it yielded no evidence of a problem in further testing. In fact, the signal disappeared again on the same line during testing at Lockheed. In addition, during these repairs it was found that one output (of thirty-two) of the 70-micron array had developed a connection to the structure of the telescope that reduced the signal and also created a risk of excess noise. This problem was in the "out of reach" portion of the cabling, so we proceeded with the knowledge that one output of each MIPS array was compromised.

Of the instrument problems, only the cable failures with MIPS could be attributed to our streamlined management approach. In the traditional management style, suppliers of critical components for flight hardware not only had to agree to a rigid set of quality requirements and then document that these requirements had been met, they also had to allow inspectors from the project to check their compliance. Suppliers dealing with NASA thus had to increase their prices, and NASA also had to pay for the inspectors. To avoid these costs, we had proceeded in many areas with "vendor certification." That is, appropriate quality requirements were levied on the suppliers, but we relied on their own internal procedures and paperwork to show that the requirements had been met. The failed 70-micron cable had been supplied with vendor certification. When we identified a fault in a connector, we obtained

X-rays of a number of other connectors in the same lot of cables. We were dismayed at what we saw inside the neat but opaque potting material protecting the connector back pins and the wires attached to them. A variety of methods had been used to make connections, and the interior looked more like a snarl of ultraminiature yarn than a neat, spaceflight-certified wiring job. None of these construction issues had been reported on the vendor certification documents, apparently because the vendor defined them as rework prior to completion of the cables rather than repairs. This semantic loophole had resulted in Ball accepting the cables and using them throughout the dewar and telescope without knowing about the possible defects hiding under the potting. Further investigations to supplement the inadequate documentation were difficult because the vendor had been taken over by another company and the people who had built the telescope cables were no longer readily available.

Every signal emanating from our instruments was carried on cables from the same vendor. However, the cabling was so intimately routed through the hardware that replacing it was not feasible without basically starting over on the assembly of SIRTF. We were not sure if new cables would be better or worse, and we had no idea where to obtain them since the original manufacturer no longer existed. Previous cryogenic missions had experienced only a few cable problems, but when that rate was corrected for the many more wires in our much more complex design, our failure rate was not necessarily any different. We pressed on, glad that so far the cables had not completely killed a science capability but knowing there was a risk of failures that would.

19

TEST AS YOU FLY

■ Before the telescope could be declared finished and delivered to Lockheed to be integrated with our spacecraft, it would have to be cooled to its intended operating temperature and run through its paces. For an "old-fashioned" cold launch telescope such as IRAS or ISO, the telescope would already be close to its expected on-orbit operating temperature inside its dewar. In essence, the dewar would also act as an environmental chamber to force the telescope and instruments to appropriate temperatures, making the necessary testing relatively easy.

The greatest weakness of the warm launch concept is the difficulty in showing that the telescope will actually cool down to the correct temperatures. In proposing the concept, Frank Low himself had felt that testing it was impossible. He had advocated skipping tests and relying on the design and careful calculations of its expected performance. However, aerospace engineers are urged to "test as you fly, fly as you test." Spaceflight engineers are far too risk-averse to accept leaving such a critical performance aspect untested.

The cooling of the warm launch telescope depends on heat disappearing into space after it leaves the outer shell surrounding the telescope. Showing that this cooling would actually occur would require putting the observatory into a vacuum chamber, the only environment where it could get cold without damage. It would have to be surrounded by a very cold shroud that absorbed its heat efficiently, since any reflection of energy back to the outer shell did not reproduce the conditions in space. Further shielding would protect the telescope within the observatory outer shell from any excess heat. The telescope would need to be shielded from the warm chamber walls to a level of one part in ten million!

Originally, it was thought that a series of cold shrouds and thermal blankets (similar to many layers of aluminum-coated sandwich wrap) could provide this level of shielding. Since we would need to include parts of the observatory inside the chamber but at room temperature (such as electronics boxes), more thermal blankets would have to trap their heat. Although all the required stages of heat rejection could be made to work on paper, it would take only a minute mistake to make the demonstration fail. Eventually, these ambitious tests were abandoned. Tim Kelly, manager of the telescope effort, described the dilemma: "An adequate and robust test approach would have cost as much as the flight hardware and been as big a development. We had to stage the testing in different temperature regimes, gluing together what came out."

We were now going to attempt the most difficult stage in this incremental test process. Ball had been preparing for more than a year to place the telescope (with the instruments and dewar) inside a large vacuum chamber called "Brutus." An elaborate system of thermal blankets would segregate areas at different temperatures. The walls of Brutus would be cooled with liquid nitrogen to reduce the heat load that had to be rejected by these blankets. Tubing would carry liquid helium to the observatory's outer shell, forcing it down to the expected on-orbit operating temperature of about 35K. We would also use liquid helium to cool the telescope and then see if it stayed at the required 6K. The telescope would be mounted on shock absorbers to isolate the system from vibration so that we could make measurements of the optical performance. The test would require $250,000 worth of liquid helium, which would be brought to Ball by tanker truck. We were used to doling out this substance from laboratory dewars that hold about 30 liters, worth about $150!

We all knew that this test was going to be fiendishly difficult, but we also knew that it was incredibly important. There was some disagreement about whether the test was important for technical reasons, since some felt a less ambitious test would verify the telescope as mission ready. But there was no question about its political significance. It would both verify the optical performance of the system, a sensitive issue after the near disaster with HST, and show that SIRTF would achieve the low temperature required to reach its advertised sensitivity. Therefore, the pretest review was abnormally large and had to be held in an auditorium off the Ball campus. The review was well attended by members of the project office, the instrument teams, and our many review teams. Ball was under great pressure to start the test quickly to win the "award fee," a monetary reward written into its contract. The combination of time pressure and the huge attendance made the review particularly ineffective at deep probing of any issues. A number of minor problems were brought to light, but Ball was given approval to proceed with the test very nearly as originally planned.

Immediately after the review, the vacuum door to Brutus was closed, the air was pumped out, and the chamber and telescope were cooled. It quickly became apparent that there was a large thermal short between the telescope and the chamber walls, and that the verification that the telescope could reach the correct operating temperatures was going to fail. The problem was elementary. As long as the helium flowed through the tubing to the telescope, it kept this tubing cold. However, when the flow was turned off to see what temperature the telescope would reach on orbit, the tubing was no longer cooled and it conducted heat inward, violating the conditions required for the test. Once this problem was encountered it was glaringly obvious, but in the rush to start the test it had been missed by the Ball engineers and also by everyone at the pretest review. The telescope team pushed ahead with the flawed test to wring all the information they could out of it.

We had no choice but to schedule a second effort. Our launch was postponed to December 2002, both to allow for the repeat test and because it had become obvious that the work at Lockheed and the preparations to operate the satellite were not going to be complete by July.

The telescope was carefully warmed up and additional cooling coils were added to solve the problem with the cryogenic tubing. In October, the Brutus chamber was pumped out and cooled again. As the telescope temperature slowly dropped, it became clear that it was not approaching the goal of 6K, but was going to level out to a much higher temperature, perhaps 15K or even 20K. There was either a serious problem with the telescope or one with the test.

Ball modified the analysis to look at the effects of changes in the temperatures as a function of changes in the test conditions. They intentionally induced different conditions, such as warming up shrouds in the chamber or turning on heaters attached to various parts of the telescope, to provide data for this analysis. Unfortunately, large cryogenic experiments adjust very slowly to new thermal conditions, so the data were all obtained by watching the rates of change rather than the end points, a procedure with limited accuracy. Thus, the analysis was complex and the results subject to significant errors. Nonetheless, Ball was able to verify many of the assumptions about the design and to show that the hardware was generally in agreement with expectations.

However, there was still a constant heat load on the telescope that could not be accounted for. Initially it appeared to be about 50 milliwatts, less than one-thousandth of the output of a typical light bulb but enough to compromise fatally the goals of SIRTF. Many felt that the problem was really a leak of outside heat into the telescope and that the test was teaching us more about the test than about SIRTF. Of course, such views were matters of belief, not engineering, and we needed the latter. We appointed a tiger team to work with Ball to interpret the test data. With substantial uncertainties, they accounted for a number of heat leaks into the telescope and derived a final estimate that there could be up to 19 milliwatts of unaccounted heat; a skeptic could place the upper bound somewhat higher.

A panel of six world-class low-temperature physicists was assembled to look over the material assembled by the tiger team and deliberate about possible causes for the behavior. In the end, all of this effort produced a perfect split in opinion. One-third of the panel felt we should repeat the test, one-third thought that the test was too difficult to be successful and we should accept what we had, and the remaining third could not decide. Tim Kelly declared that the Ball personnel were exhausted from the two tests run already, and that a third might result in a mental error that could damage the flight system. The combination of this position, the implications for our schedule and cost, and the lack of direction from the blue ribbon panel resulted in our accepting the test results as final.

Ball engineers produced a "reconciled" thermal model based on the measurements made during the Brutus test. The model appeared to confirm their previous estimates of more than five years of mission life, finding negligible change in the mission properties and expected lifetime. An implicit assumption was that the 19 milliwatts of "mystery heat" would go away on orbit. Various activities were carried out in response to concerns brought to light by either the tiger team or the blue ribbon panel, but in the end none of them suggested problems with the telescope design. Just in case, an operations team at JPL considered how the mission would be affected if we had 19 milliwatts of extra heat, and found that the primary science goals could still be achieved with a mission barely two and a half years in length.

Thus, even though it had not really been tested fully, the telescope was warmed up and removed from Brutus. After the cables that had caused problems with instrument operation during the test were replaced, the telescope was ready to ship to Lockheed to be integrated with the spacecraft.

We, along with the rest of the country, were jittery about the potential aftermath of the destruction of the World Trade Center on September 11, 2001. The telescope shipment was kept secret with regard to time and route. High-class truckers have high standards for their trucks, though, and the rig that carried SIRTF was so highly polished that it virtually glowed in the dark. Nonetheless, the telescope arrived safely in Sunnyvale on February 25, 2002, more than a year behind the original schedule in the "June deal." However, it arrived easily in time for integration with the spacecraft, which had slipped by even a little more.

20

2002: "JUST IN TIME" MANAGEMENT

■ Everyone had become immune to the Lockheed PowerPoint slides predicting success just around the corner. A management retreat was held in May 2001 to try to improve the situation. Lockheed and JPL personnel were asked to describe their goals. In areas like the overall success of SIRTF there was good overlap, but the areas of difference are interesting. Lockheed consistently emphasized financial aspects, such as award fee scores, future business, and advantageous competitive position. JPL emphasized developing new technology, impact on the public, and making a difference in a historically important enterprise. The Lockheed view was more pragmatic and less idealistic, more short term, and more difficult to convert into a vision that would inspire people working on the project to do an outstanding job. Despite efforts to put a positive face on the retreat, some participants felt that it brought out irreconcilable differences between Lockheed and its customer, JPL.

Upper-level Lockheed management became increasingly concerned about the problems. At the end of 2001, Tom Dougherty and John Straetker were put in charge of the Lockheed effort. It was a bit like changing coaches on a professional sports team in midseason. The team seemed to be playing reasonably hard and the old coach kept predicting a winning streak, but it never seemed to happen. A new expert hand at the top might implant a winning attitude. The new coaches had strong reputations within Lockheed for their ability to pull programs together.

The spacecraft software had fallen so far behind schedule that it would not be ready for a test of the full observatory—the spacecraft and telescope joined together—until June or July 2002. The long cascade of tests needed to show that the observatory was ready to be launched was going to take another eight months, so there appeared to be little chance on the traditional course to launch SIRTF anytime near the scheduled date in December 2002. Yet another launch delay would cost NASA tens of millions of dollars and damage reputations at JPL and Lockheed. The project office and the new managers worked out an approach to put the effort on track and keep it there. The plan was ready in February, a few weeks before the telescope arrived from Ball. Lockheed was to extract usable pieces of code for the observatory testing without integrating it into a full set of flight software. The programmers would finish up the flight code in parallel with our testing.

As we approached the test deadlines, software emerged and actually worked! "Just in time" management had just barely succeeded in this case, although the slow development of the software had had broad impacts across much of the observatory, from the organization of the test plan through the integration of the instruments to the ground operations development.

Tests of spaceflight hardware are supposed to be conducted in accordance with written procedures that are prepared in advance and prescribe every aspect of what the test engineers are to do. The engineers are usually focused on preparing the hardware and software required for the test, often a major task since each test may involve unique combinations of support gear that have never been used together before. Thus, the test procedures are often prepared just in time, to the point of writing up the next test while the last one is still proceeding. At Lockheed, we struggled with this cycle of preparing for tests and writing procedures. We encountered difficulties continuously, but we also made steady progress. Many problems cropped up in the testing, but none proved to be substantial at a level that would stop the project in its tracks (as had been the case with the dewar overpressure and the Brutus test). By a lot of scrambling and overtime, Lockheed was just able to compensate for the myriad delays and hold to the overall schedule.

The observatory was set up in a large clean room, where it was bathed in a diffuse light that seemed disconnected from its sources at the top of the room. The control room for testing looked out over the observatory through a row of large windows. Clean room–garbed technicians would enter from time to time and cautiously make an adjustment. SIRTF no longer appeared to be a mundane machine. It was easier to understand by imagining that the room was the temple of some strange pagan religion. The observatory served as the idol. White-garbed acolytes came with great caution and respect to make offerings. Perhaps this version of events is not as bizarre as it seems, considering the huge effort and care that had gone into the machine standing in radiant perfection on the clean room floor.

Soon there was only one hardware test to go. The observatory was removed from its temple and carted to a huge and ugly vacuum chamber. The door to the chamber was closed and the air was pumped out. The entire observatory was put through tests at the temperature extremes that represented the minimum and maximum specifications to which it had been built. The goal of these tests was to show that, in the vacuum of space, all of the controls would work correctly to maintain the temperature in the spacecraft within the range where all of the electronics would work.

Completing the spacecraft had proven to be surprisingly difficult. Areas in a mission payload like our science instruments and telescope get into development trouble because they are unlike any devices built before. The first-of-their-kind parts of a science mission are the heart of its promise, so the risks they pose are unavoidable. But we had had just as much trouble with our spacecraft as with the payload. Everyday devices like refrigerators and automobiles are reliable because millions have been built and the public has acted as a vast testing laboratory. Spacecraft like ours fall between the one-of-a-kind and mass-production categories and so are easily underestimated. Thousands of spacecraft have been built, and it is easy to get into the same frame of mind in ordering up a new one that might be appropriate in ordering a new Ford. Just pick the model sized for your needs and go down the options list checking off the features you want.

Lockheed Sunnyvale seemed to operate in the Ford production mentality, perhaps because many of the company's other programs involved building a series of identical satellites. As a corporation, Lockheed did not seem to have the inquisitive nature that is generally helpful in building the first item of its kind. As the effort was originally advertised—build a spacecraft just like previous ones and reuse the software from other missions—the production approach may have seemed appropriate. As we progressed and the uniqueness of Lockheed's contribution became apparent, however, the corporate approach did not seem to adjust.

Lockheed also had internal problems. Work on SIRTF was unattractive to employees because the project was isolated and offered no career path beyond the three to five years of the program. Eventually, a core of talented, highly committed engineers was distilled from the combined personnel pools of the Sunnyvale and Denver divisions. One highly competent Lockheed engineer from this group, who had to help clean up the mess, described our spacecraft as "possessed" because of the many poor decisions that had been built into it.

Two of the items we had checked on the spacecraft options list caused problems that were very difficult to overcome. The first was to reuse software from previous programs, and the second was the cold gas reaction control system.

Lockheed had intended to obtain much of our software by transfer from that being developed for Mars missions at Lockheed Martin in Denver. Frank Martin, who led their proposal effort, summarized the strategy: "We thought it would help win the job at JPL to promise software from the JPL-based Mars missions. There was also the usual optimism about reuse."

The heritage software seemed attractive for a number of reasons. Our mission and the Mars missions would use the same computer, and the same newly merged company was building the spacecraft for those missions. Unfortunately, the Mars missions were still under development. The benefits of using heritage software are compromised if it does not provide a stable foundation on which new developments can build. Heritage from code still under development might be described by the oxymoron "future heritage."

These problems were exacerbated by the differences in JPL's and Lockheed's views of how to organize the company's overall SIRTF effort. JPL saw two distinct responsibilities: the first to build, integrate, and test parts of the mission; and the second to provide overall project systems engineering (the discipline that takes an overview to make sure all the parts will work together). JPL pressured Lockheed to divide these tasks so that the systems engineering effort would not be perceived to have a conflict of interest in making tradeoffs that affected the rest of the Lockheed effort. When this reorganization took place, it brought new managers into the effort. The original Sunnyvale proposal team had negotiated promises for transfer of the software and its support from the code developers in Denver. These promises would have been difficult to enforce in any case, because they had been made at the upper level of management and had not been negotiated at the working level where they needed to be implemented. Since the new managers had not been participants in the proposal, they faced increased difficulties enforcing the terms with Denver. In addition, they were veterans of the historic rivalry between Denver and Sunnyvale (still with wounds from proposals lost to the Denver division), making them distrustful of the "rival" organization. For all these reasons they chose to initiate an independent software program in Sunnyvale. The shift away from Denver removed any realistic possibility for the necessary close support to allow effective incorporation of the heritage code. However, the funding was still at a level requiring the heritage approach, and nominally we continued in that direction.

Without the strictest discipline, heritage can undermine a rigorous approach to software development in which a complete set of requirements is written, a software architecture is built around them, and finally code is written to carry out each task. With heritage software, initially it often appears that only some minor additional feature is necessary. Only later is it realized that yet another feature is needed, and another after that, and so forth. Even though the initial code may have been tightly designed for its purpose, with this kind of growth, key aspects may not keep up. In sum, the piecemeal growth of heritage code invites underestimates of the ultimate effort required, and our group was seriously overcommitted without being aware of it.

The nominal similarity of our hardware to that of the Mars missions further encouraged overconfidence. Although this analogy was valid in areas like telecommunications and management of electrical power, SIRTF also posed many problems not represented in a Mars mission. These areas included pointing a telescope accurately, preventing the telescope from pointing to forbidden regions of the sky, and managing huge amounts of scientific data. Because of the many new requirements for SIRTF, much of the spacecraft's electronics was also largely new, requiring definition of many new interfaces so that it and the software would work well together.

These issues are supposed to be dealt with through a systems engineering overview that insists on development and documentation of requirements and negotiates interfaces with other areas in the project. Lockheed was ill-prepared for this responsibility because of the strategy to hold the spacecraft spending back to fit the funding profile that NASA had made available. At the time when the spacecraft Preliminary Design Review (PDR) was supposed to endorse their overall engineering approach, Lockheed had only a small staff assigned to the project; funding to increase the effort to a more adequate level did not become available until a couple of months later. They were already behind schedule at the PDR—there was no overall observatory requirements document, for example. Given the very fast schedule that the project was operating on as part of the strategy to minimize the overall budget, Lockheed was always scrambling to catch up. Nick Vadlamudi, who joined the project at the time of the PDR as lead systems engineer, reflected that it might have been better to stop altogether and establish the missing systems engineering foundation before continuing. Doing so, however, would have made it apparent relatively early that Lockheed was not going to be able to hold to the overall schedule. Much, much later, Gary Pace, who had been brought in to help manage the software effort, drolly described the tardiness of the system requirements documentation: "The requirements definition is coming along pretty well—we should have it done by the time we have written the code."

The rest of us unwittingly contributed to the problems. Our years of deprivation had made us alert to superfluous complexities in the hardware; however, software complexity can grow in subtle ways. New software has no mass and requires no space, so improvements can be made nearly at any time in a satisfactory minimal solution. Thus, to achieve gains in the observing efficiency we added complex new software to speed up the settling of the telescope on a new target; an onboard scheduling tool that would select the next observation from a list, so it would fit within the available time and minimize the requirements to repoint the telescope (this feature was eventually abandoned in a very late attempt to simplify the software); and a standby mode for the spacecraft, so if something went wrong in a minor way we could resume observing more quickly than if the spacecraft had "safed." To some, these enhancements made the code take on aspects of a research project rather than a tightly controlled development with hard deadlines for delivery, rigid requirements, and extensive testing of software modules as soon as they were good enough rather than trying to make them better.

All of these issues came to a head at the time of the spacecraft CDR in September 1998, when we should have had the software task firmly defined with a good understanding of its cost. Instead, the estimated lines of code were steadily growing, and the cost estimate was based on rough models and extrapolation. The fault protection (code to detect and deal with problems in the spacecraft) had almost been forgotten; it was estimated to be at no more than 10 percent of its eventual scope. At our next management team meeting, we learned that Lockheed had hurriedly assigned seventeen people to the effort, without even consulting JPL.

Larry Simmons [to spacecraft manager Milt Whitten]: Do you need budget authority to do this?

Whitten: I can't wait. We're moving out, spending at a higher rate than allocated. Fire me if you want, but that's what I've got to do.

Larry: Not until you're finished.

Along with the increased manpower came a request for a big increase in budget, which was quickly granted.

Although the original plan was modified dramatically as the work was better understood, both JPL and Lockheed were locked into the "June deal" budget. Neither party wanted to talk about the condition of the emperor's clothes. In addition, the tendency of software to cause schedule problems for NASA had risen to the level of an expectation. Programs had learned that they could recover from inadequate requirements definition through extensive testing and rewriting, an approach that is more feasible with software than hardware but still generally results in schedule problems and cost growth.

When disaster appeared to be unavoidable, the effort was drastically reorganized and nearly one hundred programmers were assigned to it—six times the seventeen that had been thought adequate a couple of years earlier. At this point, huge inefficiencies had to be accepted. Each new programmer has a period of orientation, and much of the orientation information must be drawn from other programmers who could otherwise be producing or testing code. Some rough estimates illustrate the nature of the problem. Assuming a typical rate of code generation and no benefit from the software heritage, the final scope of software written for SIRTF should have been about three times the original estimate. Thus, there was a substantial initial underappreciation of the task. However, the cost to the project was about five times the original estimate. That is, the inefficiencies associated with the chaotic early development and the huge increase in staff to compensate resulted in major additional costs. The continuous threat to the testing program due to slow delivery of the software added costs in other areas as well.

We checked the box on the spacecraft options list again for a cold gas reaction control system (RCS). The logic was simple. The telescope could be pointed to different parts of the sky by pushing against torques from a system of rapidly spinning reaction wheels, like giant gyroscopes. Corrections were also needed for external influences, such as the pressure from the light of the sun striking the array of solar power cells. These pushes would gradually increase the spin rate of the reaction wheels until they had to be slowed down by pushing against something else. The something else was to be provided by an external thruster system.

We did not want conventional thrusters for this task. Some small portion of their exhaust might scatter back toward the telescope, where it was likely to freeze out and degrade the performance of both the telescope and the surfaces that were supposed to keep it cold by radiating heat into space. The problem could be avoided by using thrusters that operated on stored gas of a type selected to avoid the contamination problem. Cold gas thrusters are commonly used in spacecraft—for example, to bring them to the right attitude immediately after launch. Perfect solution, meets requirements, has heritage, check.

Only after we were firmly committed to the cold gas RCS did problems begin to appear. A cold gas RCS consists of a reservoir for highly compressed gas, a system of tubes to carry this gas to various points around the spacecraft, valves to control the flow through the tubes, and a special valve at the end of each piece of tubing that acts as the thruster. When the thruster valves are activated, short bursts of gas are released to steer the spacecraft. If the system was to operate for the full five-year mission, the thruster valves would have to close tightly when we were not unloading the reaction wheels so that gas would not escape and bleed down the supply. It was known that thruster valves tend to leak a bit no matter what, so the tubing for the gas was made small, and "low-pressure latch valves" were placed upstream of the thrusters. After a burst of thruster activity, the tubing to the device could be sealed off with these additional valves and there would be only a small amount of trapped gas to bleed out. The small tubing and other requirements necessitated very small thruster valves, at the limit of what could be produced.

Right from the beginning there were problems. A couple of thruster valves failed their leak tests. It appeared that small particles in the tubes feeding them were getting blown up onto their valve seats and sticking there. When the valves closed, they not seal properly. Filters had been put in the tubing runs to catch any particles, but they were not getting the job done. Inspection of the tubes indicated that they were not as clean as they should have been, and that the filters themselves were shedding particles. Months were spent improving the assembly procedures for the tubing and improving the filters. Yet the problem persisted, and in peculiar ways: the leaks changed size, came and went.

One of Dougherty and Straetker's first efforts after taking control of the Lockheed work had been to get the thruster system assembled on the spacecraft. Given the symbolic importance of this first key event in their reign, they carefully developed a plan to work around the assembly problems, had technicians practice the necessary welding steps, and when all was in place gave the order to proceed. Embarrassingly, the welding just did not work on the tubing that had been purchased for flight. The mystery was finally solved when it was discovered that the supplier had shipped nickel tubing instead of stainless steel, making the carefully developed welding procedures inappropriate. Although this setback was quickly overcome, it was symptomatic.

Once the system was together, most of the thruster valves behaved correctly, but two of them were never the same from one test to the next. Others showed hints of similar unpredictable behavior. Lockheed's initial recommendation was just to replace the valves affected and continue with the program. Such an approach would be appropriate on an assembly line where one was confident that the part involved was defective in a unique way. Because our valves were custom-made to our specifications, we had to allow the possibility that they all suffered from a general weakness. In addition, the replacement process was going to be ugly. It called for cutting the tubing to remove the thruster and arc welding a new section on the exposed end. All of the flight electronics was mounted in the spacecraft, and millions of dollars and many months had been invested in testing the combination. The prospect of arc welding on the same structure, releasing thousands of volts within inches of the electronics boxes, gave many of us nightmares.

Events quickly forced us to abandon the "single part, unique failure" theory and pursue the underlying cause of the problem. At one point the entire observatory was put in a heavily walled room and bombarded with intense sound to simulate the noise of the rocket when it was launched. At the end of this test, all the thrusters had stopped leaking. As testing continued, the leaking came back. We simply could not be confident that the system would work, particularly after a newly manufactured spare valve suddenly developed a large leak, and in a way that implied that the particles had come from inside the valve itself, not from the tubing feeding it. The leaking spare valve implied that we did not yet know enough to get replacement valves that were better than the ones we already had.

The progress toward completing the spacecraft slowed as we struggled to find a solution. As Straetker later said, "When we came onboard, we worked hard to put thirty days of reserve into the schedule, and then in the first thirty days we used twenty-five of them." Because the RCS problem threatened to delay completion of the entire observatory, we had to spend money pursuing multiple solutions until the right one became obvious. A team practiced welding techniques off the spacecraft with instruments monitoring the voltages on the tubing to show that it could be done without destroying our electronics. Another team developed new assembly methods that did not require welding at all. Yet another team assessed the consequences of accepting the situation and using other valves located farther up the line to shut off the leaks. Meanwhile, the Science Center explored ways to keep the spacecraft under control if the gas did run out by carefully selecting research targets placed so that the pushes on the reaction wheels canceled and never built up to the point that the thrusters were needed.

Even if we found a way to avoid replacing the valves, we had to understand the problem thoroughly to be sure we could handle it. A valve that had been rejected early in the spacecraft build was cycled many times and then cut apart and inspected. The pictures showed material coming off the valve cylinder walls and piling up into a ridge where the valve seat moved. It was likely that *all* of the valves welded into our spacecraft were defective in this way, along with the replacement valves we had been purchasing. A detailed investigation was conducted at the manufacturer's plant. A "qualification" unit had been built and tested to many times our needs, and it had worked perfectly, giving confidence that the flight units would be satisfactory if they were made in the same way. But this valve had been built as a prototype by an exceptionally skilled technician under no time pressure. Once the concept had been proven, the flight valves had been built in a production environment, with different technicians working two shifts and with considerable pressure to hold schedule. One of the first lessons in critical spaceflight construction is that there are nuances that are not communicated in written procedures. Super technicians are super because of their instinctive mastery of these nuances; others cannot easily reproduce their results. Something—number or placement of welds, details of welding technique, who knows—had been different for the flight build.

We now knew the cause of the bad behavior of the valves, but replacing them so late in the program was going to cause many months of delay while new ones were manufactured and welded onto the spacecraft tubing. Our extensive explorations of alternative solutions indicated that we could manage leaks. We redefined the low-pressure latch valves to be our first line of approach for retaining the reaction control gas. In this new role, these valves would be operated more often than in the original plan. Additional tests showed that they should be reliable.

With a collective sigh of relief, we decided we could manage with the RCS we had, despite its long history of problems. Nonetheless, this seemingly "standard" option had turned out to be a lemon, and it had cost us huge amounts of time and money.

With the completion of the thermal/vac test, the progress on our software, and resolution of the issues with the RCS, it was obvious to everyone that we were on the road to a launch pad in Florida. The plans put in place at Lockheed in February to keep the effort there on schedule and within budget had worked amazingly well. Every time there had been a test delay or a problem, some approach had been found to reschedule the activities and miraculously make up just enough time. Although initially we had eaten into the schedule reserve alarmingly rapidly, we leveled out with about ten days of reserve left and held there with no additional slippage for nine months, until the job was nearly done.

Straetker modestly claimed that the success had been easy: "Once the entire observatory was in place, everyone could see the focus of all the effort and they did what was necessary. In addition, we had been able to save about $4 million, and we used it as a reserve to pay for additional work when we needed it."

At our October 2002 management team meeting we discussed the final steps toward shipping the observatory. The level of the project reviews had been increasing rapidly. However, many of them were relatively superficial, often bringing up issues gleaned from our own risk list. This type of review had far greater potential for micromanagement than for catching some problem we had missed. Everyone involved in any way wanted to show that they took a personal interest in our success, but not necessarily in a way that would help us succeed. It was going to be largely up to us to catch any real problems.

Gallagher told the assembled team: "I value far more the opinions of the people around this table on whether we're ready to launch than I do the review boards. We control the information they get and they seldom discover anything we didn't know. I'm not concerned about the dates for these reviews, but I want to be sure everyone on the team agrees we are ready." Irace added: "We don't want to be like the wedding is scheduled, and no matter how we feel we've got to be at the church on time. We need to know now if everyone can support the schedule we have up there, and I hear serious concern up at that end of the table [from the Science Center] about our readiness to operate the observatory. If we know it now, we can adjust the schedule."

As the Science Center representatives had pointed out, there was still one more bill to pay, this time for the decision to proceed with testing the observatory using software that was not completed to full flight standards. As improved versions of software became available, each needed to go through "regression testing" to show that all the features that had worked before were still in order. A necessary side effect of this approach was that there would be less time to operate on the software once it was "done" and therefore a larger chance of serious problems not being caught until after launch. A swarm of software problems harassed us, and as each was resolved it seemed that another was uncovered. The SIRTF deep space orbit required that the observatory work nearly perfectly because it would have to operate without continuous monitoring and potential intervention from Earth. By mid-October it had become apparent that a new version of the software was needed to fix a couple of dozen nasty problems. Gallagher had described the size of the problem at our management meeting at the beginning of the month: "The operational readiness and the set two tests are not going to stop the launch in January. They may delay it a couple of weeks, but that's it. When it comes to launch delays, this is it."

Gallagher composed the letter to inform NASA Headquarters of the need for a new software version and the resulting launch delay, but decided to sleep on it and send it the following day. That morning, it was announced that there had been an accident mounting the Global Positioning System satellite (GPS) payload on the rocket ahead of ours on the launch pad. A crane operator had misinterpreted a "thumbs up" signal and lifted when the satellite was already bolted to the rocket, and the resulting stresses had caused damage that would take time to repair. We would be forced to accept a launch delay while the new parts were manufactured. A new launch date of January 27 (soon adjusted to January 29) was imposed. Thanks to that crane operator, the launch delay was attributed to the assembly accident and not to us.

On November 7 we assembled for what we thought would be our last prelaunch quarterly review. The room with the JPL directors' portraits had been preempted by a high-level JPL management review, and we were squeezed into a smaller one. By now the reviews had become a continuous blur; it was reported that we had been through 199 of them—1 per week—since our CDR about four years earlier. The rate was rising toward one every other day up to launch. Every review had the potential to load major tasks onto our already stressed team.

Despite the pressures, we were making rapid progress. Gallagher encouraged us to keep our personnel well rested so we did not make a mistake: "We have a fabulous observatory sitting up there in Sunnyvale. The biggest risk is that we do something stupid and damage it."

21

1999–2002: HOW DO WE DRIVE THIS THING?

■ Before we could launch it, we had to demonstrate that we knew how to drive our complex machine. Imagine trying to learn to drive a car by sitting in it on the showroom floor and reading the owner's manual and booklets from the licensing bureau. To complete the analogy, imagine that the car is the first of its type ever built, and that the authors of the manual had been given no opportunity to drive it! Coping with a challenge such as this requires a lot of planning and practicing, anticipating all the things that might happen during launch and on-orbit operation. These processes are generally lumped under "mission operations." In addition, we needed "data analysis" capability to understand the data we got back from our instruments. Once we knew that SIRTF was working, the Science Center also had to be prepared to accept and review proposals from astronomers who wanted to use our machine, and to reduce and calibrate the data for them.

The June deal had come at a unique time for mission operations at JPL. The lab had been working to reduce the costs for this process by developing a system that was general enough to be used repeatedly. This multimission system had been implemented in part on Magellan (to Venus) and Mars Observer, and in full on Cassini, which spent only a tiny portion of its billion-dollar budget on mission operations.

Mars Pathfinder had provided another example of very inexpensive operations. At the time we made our initial plans, Pathfinder mission planning and training were nearly complete, drawing on the JPL multimission approach. We took tours to see how the mission-specific software had been developed. In the middle of a large building was an adult-sized sandbox with real sand and rocks arrayed to look like the Viking pictures of the Mars surface. Offices of the programmers for the mission surrounded the sandbox, and there were windows all around so it could be easily viewed. In the middle was a spare Mars Pathfinder rover. The setup looked like a playground for adults, with a very fancy radio-controlled car as the main action. The programmers could even watch the behavior of their software from their office doors. Although the success of the approach had not been demonstrated, the Pathfinder sandbox became the subconscious model for our mission operations, a bit as the Mars program software provided "future heritage" for our software.

The analogies with our situation were flawed in both cases. Cassini had received a large amount of additional software support from the budget for the multimission operations system. Its team was allowed to use this money for their operations needs so long as the software was sufficiently general to be used by others as well. In addition, they had a seven-year coast in space to reach Saturn. They launched without a system to reduce, calibrate, and distribute the science data and developed it during this long period of reduced activity. Pathfinder also launched without all of its operations software in place and had its last operations readiness exercise only shortly before it arrived at Mars after nearly a year of coasting in interplanetary space. We had no access to "hidden" funding, and all of our system had to begin working immediately upon launch. Nonetheless, following the June 1997 deal, our operations leader made a valiant effort to make the greatest possible use of the multimission operations capabilities to hold our costs to an unprecedented low level for a major multiuser astronomy mission.

We began to depart from this direction, however, when a few months later NASA Headquarters decided that the SIRTF Science Center (SSC) would be at the Infrared Processing and Analysis Center (IPAC) on the Cal Tech campus, about a twenty-minute drive from JPL. The growing realization that the low-budget multimission approach was not going to be adequate, combined with the independent selection of the SSC and its physical separation from JPL, plus a history of rivalry between campus and lab, all permitted the SSC to operate in a relatively independent fashion.

Bill Green was appointed manager of the SSC. He had worked for years in the JPL multimission program and knew it well. It was clear to him that an astronomical observatory differed in important ways from a planetary mission like Cassini. Our telescope had strict avoidance constraints that had to be honored by the operations. We had a complex payload of three science instruments, each of which required slightly different ways of operating the satellite while gathering data. A high volume of data would flow from these instruments starting shortly after launch. There would be a large astronomical community vying to use SIRTF, proposing new programs to which the mission timeline could not be adapted in advance.

Clearly, the multimission software was going to need extensive modification. With its developers rooted entirely in planetary exploration missions, Green was concerned about getting the priority that would be needed for such changes. Therefore, the SSC took software designed for the Hubble Space Telescope as a basis instead, in the belief that the underlying issues would be similar because HST was also an astronomical mission. This package was renamed and extensively rewritten to take the science programs and convert them into the sequences of commands needed to operate SIRTF. The SSC also set out to develop the systems to reduce the science data and to monitor the health of the instruments.

Other operations tasks remained the responsibility of JPL: preparing the commands to be sent to the spacecraft, and collecting and formatting the raw data from the Deep Space Network of radio antennae that would receive the telemetry from the satellite. The mission operations group at JPL was limited to about twenty-five people who were spread over almost as many areas. This scale of operation might seem plausible by analogy with the Pathfinder effort, but this comparison ignores the "diseconomies of scale" that are characteristic of complex technical endeavors. As the size of a project grows, it must be divided among an increasing number of people, and the number of potential interactions grows rapidly. For example, with two people working on a task, there is only one interaction, the one between them. With ten people, there are potentially forty-five different interactions among the members of the group—an increase by a factor of forty-five for a fivefold increase in the number of workers. Thus, scaling in a simple way from a small effort like Pathfinder to a much larger one like SIRTF is likely to arrive at an underestimate of the management complexities.

The situation came to a head in the Mission Operations System CDR held

September 19–21, 2000. Although many pieces of the required operational system had received attention, they had not been integrated into a functioning system. Thus, the review board found that

> no integrated schedule showing all the "work to launch" . . . was shown. . . . The STL [Systems Test Lab] is a key tool for operations both in the development as well as in the on-orbit phases of the mission. The fidelity of the STL to support operations development and on-orbit anomaly investigation was not discussed. The schedule for the use of the STL was not discussed. The validation of the STL against the flight hardware was not presented. . . . The . . . staff allocated to training was inadequate. . . . The process for capturing, implementing, and validating the flight rules and constraints was not fully described.

Most important, however, the board reported "a clear need for an operations manager with past experience in developing operations products through the final spacecraft integration phase, and with experience preparing an operations team in the use of those products on-orbit. Such a manager should be deeply involved in the spacecraft testing as part of the operations product development. . . . A single hard-driving operations manager is also needed to integrate, prioritize, and manage the work leading to launch."

Gallagher eventually found that manager in the person of Bob Wilson, who joined the project on February 15, 2001, nearly six months after we had failed the operations review. Wilson had been a manager in the Telecommunications branch in JPL and hence had a network of contacts in mission operations at the lab. He was a lucky find. His experience gave him strong views on the way to solve our operations problems, but he expressed his opinions in an extremely polite, almost gentle (if immovable), and patient fashion. This combination allowed the operations team to be augmented and trained with a minimum of friction.

He had to move very fast to recover from the review failure and the lost time since then. First he established a red team of experienced operations personnel to supplement those already in place. These people were drawn from many projects and could not be transferred to SIRTF permanently. Given the high priority within JPL to solve the problems with SIRTF, however, Wilson was able to use them to jump-start the SIRTF operations and to mentor new people who were brought in to take over. By the time this process had run its course, the original staff of about twenty-five had tripled in size. As usual, the cost of recovering from getting far behind considerably exceeded what we would have paid if the operations effort had proceeded smoothly from the beginning.

Operations and flight software are intimately intertwined, so two of our problem areas overlapped. Lockheed Sunnyvale's dual responsibilities for software and spacecraft monitoring during operations threatened to compound the difficulties. The Sunnyvale management was so overwhelmed by the issues with flight software that they had done little to prepare to support operations. Lockheed Denver had a strong reputation for operating spacecraft, and its capabilities were underutilized. Sunnyvale was unwilling to give up its original role, but given its overall performance, JPL was anxious to transfer any challenging new work elsewhere. The issue was forced by having a side-by-side review, which dramatized the lack of preparation in Sunnyvale and the good understanding of the process at Denver. The operations role was transferred to Denver soon after the review.

Although the SSC had been given passing grades in both operations reviews, it also had a major challenge. Its initial funding had been colored by the general optimism that had prevailed in the June deal. At one point, the project had even taken a budget of about $35 million a year for postlaunch operations to NASA Headquarters and been told by Ed Weiler that it was two times too small and that he was going to force a $70 million budget on us.

Soon, even this budget began to look thin. In order to gain as much efficiency as possible, we took steps to:

use an efficient planning concept based on observation templates borrowed from the ISO mission;

operate only one instrument at a time to make planning and scheduling easier, since the observing could be divided into weeklong campaigns centered on one instrument;

assign instrument support teams (ISTs) at the SSC to work exclusively and closely with each instrument team, developing common interests and collegial relations;

have the ISTs work with the instrument teams to test command sequences early while the instruments were being tested, giving the SSC a head start in understanding how to operate the instruments; and have the instrument teams test various versions of the SSC-provided science planning software as part of the preparation of their guaranteed science programs.

The last three items encouraged optimal use of the talent within the overall SIRTF team. We needed to avoid the friction that had sometimes characterized relations between instrument teams and the (Hubble) Space Telescope Science Institute. The institute had established parallel efforts to duplicate the work of instrument teams in data reduction, calibration, and instrument operation. The two groups had fundamentally differing motivations: institute personnel felt rewarded when they found a problem with an instrument, while instrument team members felt rewarded only when they *solved* such a problem. Furthermore, institute personnel were expected to implement secure, conservative solutions to any problems, and thus tended to react more slowly than the instrument teams.

Eventually, the operations teams and the SSC had put all the operations tools in place. All the hardware tests had been completed. A series of final tests was scheduled to demonstrate that we really knew how to drive the satellite. We had painfully negotiated an "incompressible test list" that had to run flawlessly before we could call the job complete. Unfortunately, the tests continued to reveal problems, and of a very broad variety. Instruments suspended in the middle of their operation, timing departed from the plans, data were lost. There was a mad struggle to find the causes of these problems, leading to a rapid succession of releases in the flight software for the spacecraft and of modifications to the command sequences. The struggle lasted throughout autumn 2002.

22
COMPLETING SIRTF

■ Completion of the entire observatory was to be marked by a preship review held at Lockheed on November 21, 2002. Contrary to usual practice, there was no plan for a party after the review. In fact, the organization of the review had been a scramble, as indicated by the notification of a precise time and place only the afternoon before the event.

Gallagher kicked off the review:

We just had a successful operations readiness review [the preceding two days]. We were told that if we continue at the same pace, we will pass for launch. . . . Next chart please.

Computer operator: Actually, we don't have any charts from you.

Gallagher [with a nervous laugh]: Well, you know I'm never at a loss for words [long, somewhat awkward pause]. There's been an event with the reaction control system that is quite serious. It has the potential to impact this review. I've asked John [Straetker] to discuss it.

Straetker: Two weeks ago a malfunction in the test set potentially caused 400 high-pressure latch valve and 800 low-pressure latch valve cycles in 9.5 minutes. Then a week later a race condition between flight software and fault protection cycled one of the high-pressure latch valves

14,750 times and the other 780 times. It only stopped because a microswitch failed to open in the second valve and called attention to the problem. The valves are rated for 2,000 cycles. We are carrying out a stress and fatigue analysis and are planning to use the testbed to put 30,000 cycles on another valve to requalify the flight unit. That testing will take two weeks, and we will need an additional three to five days of pressure testing on the observatory. We also have to track down the microswitch failure. If we have to replace it, it is embedded in the valve and we would have to disassemble the observatory. This issue will require us to move off our launch date of January 29.

It was fortunate that no party had been scheduled; it would have been more of a wake than a celebration.

The rapid reversal in our prospects was stunning. The first problem had actually occurred on November 7, just before Gallagher said, "We have a fabulous observatory . . . the biggest risk is that we do something stupid and damage it." The Lockheed test crew had been puzzled by anomalous behavior on the spacecraft at that time but had not followed up to identify the cause. The second event had been much more serious. The system had tried to switch from one side of its electronics to the backup side, but the software did not handle this step cleanly and the two sets of electronics had ended up arguing about which was in control. Although the fault protection had been programmed to stop commanding after three tries, as a result of the argument an infinite series of commands had been issued in a kind of death spiral.

Certain types of high-pressure latch valve failure could cost us the mission. Because the valve had been pushed far beyond its rating, we were going to have to do additional tests and analyses to regain confidence in it. If the valve failed this scrutiny, we would have to disassemble the spacecraft to replace it. We also had to be sure there was no possibility of a similar death spiral getting established on orbit, since it could end the mission. Lockheed personnel argued that this was "just" a procedural issue; it could not happen on orbit because the relevant software test module was scheduled to be decommissioned with the next release of the flight code. However, the module was to remain in hibernation in the flight software, and we knew of missions that had been killed by accidental activation of such modules.

The review proceeded while Gallagher and the attendees from NASA Headquarters spent most of their time outside the room discussing the consequences of the news. Gallagher was pressured to decide whether we could still make the January 29 launch, because the repair of the GPS satellite had gone slowly and there was now a conflict for the launch pad on that date. He could only guess at the impact of overexercising the valves, of course, but ultimately, he stated that launching on that date was still possible. It was decided that NASA would take a gamble and pay the $2 million to move the GPS launch to another pad, clearing the way for us should we manage to hold schedule.

While this drama played out in the corridor, the review droned on. There was nothing dramatically wrong with the material presented. It was painfully obvious that we were not ready to ship, though, and that the month until the scheduled ship date of December 18 was going to be frantic even without the new task of requalifying the latch valve. Although most of the mission elements had completed testing and appeared to be in reasonably good condition, a huge amount of paperwork had to be completed to document the results.

The paper needed to launch a space mission might get it pretty close to space if it was just piled under the rocket. Each mission goes through a "verification" process, in which it must be shown that every requirement that had been set for the hardware has been satisfied or that any deviations are acceptable. The observatory testing had been completed last, and under the pressure to maintain schedule a lot of the necessary document writing had been postponed. Consequently, of eleven hundred requirements we had closed out only nine hundred. In our *best* month previously we had managed to close out only two hundred, and we now had less than a month to complete a similar number (including some difficult ones that had been postponed). Even the plan showed us barely making it, and we had been steadily falling behind plan for the last few months. In addition, every discrepancy in the requirements verification had to be recorded on waivers or "engineering change requests" and approved by any affected parties so that there would be no surprises on orbit. We were hopelessly behind on this approval cycle, and had been for years. Finally, everything that happens unexpectedly during testing must be written into a "test anomaly report" (TAR), and all the TARs must be accounted for prior to shipment. As we continued the test program, we were generating new TARs as fast as we could close the old ones. Significant TARs are upgraded to "discrepancy reports" (DRs), and these too were being closed only slightly faster than new ones appeared. A simple extrapolation suggested that it would take many months to finish all of this paperwork.

More worrisome, the only way to finish at all appeared to be to stop testing; the trends made it clear that our observatory was still not as fully understood as it would have to be when it was in space. Although reducing this mountain of paperwork was extremely tedious, it had to be done to ensure that the observatory was as close to fault-free as human beings could make it.

There was another significant problem. Despite the attempt to finish off the spacecraft software with the last release, a new set of software problem reports had already sprung up. On-orbit software fixes are always risky, so launching with a long list of bugs is not desirable. In addition, the lack of stability in the software raised the possibility of an undiscovered bug lurking within that would surface and cause serious difficulties. Although some of us might have quarreled with the rating, none could disagree with the sentiment when the red review team described the software risk to the mission as high. After the review, SIRTF Program Executive LaPiana stated privately: "No matter what Dave [Gallagher] and Larry [Simmons] think, we're not going to let them launch it until we're sure the software is OK."

The investment to make the launch pad available, the huge pile of paper to be processed, the need to requalify the high-pressure latch valve, and the demand to get the errors out of the software all combined to make our schedule pressure immense.

The next week I was in England with Larry Simmons on a trip associated with another project. On the way to dinner Monday evening, he began singing, "It's starting to look like April . . ." The air force had declared the GPS launch, which had already been delayed six times, a pressing national priority and had refused to move off the launch pad. The repair of the damage to their rocket was going slowly, and there was no plausible launch window for us prior to our deadline of March 8. We were going to have to wait until the next period we could get into the right orbit, which opened in mid-April.

One of our November reviewers appended a note in early December to his comments on our readiness: "Since there is now a 'God Sent' launch delay, much of my concerns regarding the Project's ability to complete all required tasks prior to launch could be mitigated. The Project should immediately prioritize the work to go and have all teams adhere to this plan."

This process began in a project teleconference. We would have a launch window from April 15 through the first week in May, after which the launch pad was needed for a Mars Exploration Rover launch. Gallagher wanted to use the extra time to improve our confidence that SIRTF was ready to launch. He noted:

> There will be some key thrusts: reestablishing confidence in our flight software and fault protection, and resolving the issues with the reaction control system. One of the things you will see from me is an emphasis on safety, so you will see increased effort on safe modes and things like that. I want everyone to understand that I appreciate the efforts they have put in. This is not a slap in the face—it is not related to flight software or latch valves; it would have happened no matter what we had done. Since we had told headquarters we could support a January 29 launch, the delay is not attributed to us. We will receive additional funding for the delay. We can also use Phase E funding for some of this. I want everyone to take the Thanksgiving vacation to be rested, and then we will take a week to replan our activities. We also have some good news about the latch valve, and I'd like John to describe it.

Straetker spoke next:

> We put together an analysis and test plan. On Monday night we brought the observatory reaction control system up to one-fourth burst pressure to see if we developed leaks due to these two anomalies. There were no leaks. We then let the system pressurize at 1650 [pounds per square inch] so we could watch it over the next couple of weeks. As of today there is no evidence of leaks. Also, an analysis from the vendor indicates that there is no damage to the valves. We will do an independent analysis and then audit them against each other. On the testbed in Denver, we are preparing for a requalification. We had a pretest review and are ready to start tests. When we are done with these steps, we will be able to say we have a system that is good to go.

The next Thursday we gathered around our speakerphones to discuss the new plans. Gallagher broadcast a hand-printed list over our conference video systems. The list clearly had been prepared with a lot of review and reconsideration, since it included a variety of printing styles and had a number of corrections. Reading from the list, he said that he wanted to emphasize five things:

1. Establish confidence in the flight software through additional walkthroughs, reviews, and testing.
2. Transition to an operations mentality. "We will do what I am calling

a virtual launch," he said, "where we transition the flight software per the schedule to Lockheed Denver as planned for the January launch" [Denver was going to have the responsibilities for operating the observatory once it was in space].

3. Be ready for launch with LARGE margin.
4. Health and safety emphasis.
5. Training.

At the bottom were three additional guiding principles:

1. Clear all paper ASAP (per 01/29 launch plan).
2. Do a few things extremely well. "We want to do the work that was planned," he said, "not load up with new tasks. The operations readiness review told us that we were on the right path but that it was going to be challenging to get everything done on time. Now we can fix that."
3. Improve the data analysis effort. [This was a reaction to the tendency to conduct tests, take a superficial look at the data, and move on without any in-depth analysis. As a result of this pattern, the overcycling of the high-pressure latch valve had not been detected at Lockheed Sunnyvale, but had been found only when the data were looked at in more depth in Denver.]

We turned to a draft schedule that showed two more releases of the software. Straetker explained: "There was a whole slew of [software problem reports] against [release] 6.0, and there are six to eight against 6.1. We will never load 6.0 into the spacecraft." Gallagher added: "There are no known fixes scheduled for 6.2, but we need it as a contingency in case new issues come up." (We had gotten to release 6.2.1 by early February.) The schedule showed these releases shoehorned in around a long holiday break, December 21 through January 5. We spent the rest of the call struggling with the lack of time in the schedule. Between the new software builds and the accompanying testing; half a month of vacations for Thanksgiving, Christmas, and New Year; and a reserve ("be ready . . . with LARGE margin"), work had filled the two and a half months of schedule relief like air rushing into a vacuum.

In addition to the software effort at Lockheed Sunnyvale, a crack team was assembled at Lockheed Denver under Rick Kasuda to work with the project to review the software. This effort was kicked off December 15 and ran at peak effort (about fifteen people) for two months. Kasuda placed the review in context with the three phases in writing and testing software: (1) development (requirements definition, coding, integration and test); (2) verification (formal testing of the code for acceptance, use in the overall observatory test program); and (3) validation (stress testing, running software sequences in a spacelike manner, testing the fault protection). His assessment was: "The program has done a really good job at putting a great validation plan in place. However, there is a little black cloud over some of the earlier steps, and they can result in little bugs that cause a variety of problems." He explained: "If the requirements and coding are not done well, there can be problems in areas like extreme situations that are not probed in the later validation because they are entered only rarely, but they can occur in the long operations on orbit."

The team had carried out a series of actions to look for such problems. They developed a custom build of the code with all the available checks of software states enabled; they also ran the code through two C compilers with all the compiler code check features enabled. The software itself had been written with "technical performance metrics"—internal checks of the flow of code at five hundred to a thousand points in its operation—and they reviewed this information. This review discovered a problem in a dynamically addressed area of scratch memory (the "heap"), which had the potential to become disorganized. The team audited the documentation from the compiler vendors to see if the vendors had found problems in their products after the relevant parts of our code had been completed. They looked for areas of code with an exceptionally high rate of problems. This search led to the mass memory card, for which 180 software problem reports had been written. A nearly full memory card seemed to trigger other problems in the system. They took an in-depth look at the fault protection code, concluding: "Fault protection is a very touchy subject and you can't say it will protect the spacecraft for all time against all issues, but . . . we think it will do the job SIRTF needs." They also searched for latent pieces of code that could create another event like the overcycling of the high-pressure latch valve. And they reviewed the actions taken in response to other reviews, to be sure that previously noted deficiencies had been taken care of appropriately. The intense scrutiny produced about 25 new software problem reports. Two of them—the usage of the heap and the problems with a full mass memory card—had the potential for substantial impacts on our mission.

Everything else was coming together. We were even beginning to make progress on the mound of paper. The engineers working in this area would send out items to be checked and approved by e-mail, and then start calling a day or two later to nag us to reply; this "system" was starting to work.

Gallagher and Irace were worried that the preship review panel would have very high standards for us, given that we had not impressed them on November 21. They held an "internal preship review" at Lockheed on January 30. It showed that a lot of progress had been made in the intervening two months, but there were still a few rough spots.

> *Columbia* Destroyed during Re-entry
>
> Seven astronauts and NASA's first space shuttle were presumed lost over the skies of Texas on Saturday as *Columbia* was re-entering Earth's atmosphere to wrap up a 16-day science mission. A Texas television station captured video of the shuttle breaking up into multiple pieces as the vehicle flew at an altitude of some 200,000 feet and was moving about 12,500 mph. (Headline in *Space.com*, February 1, 2003)

Fortunately for us, as a result of the *Challenger* disaster seventeen years earlier and the resulting doctrine of a "mixed fleet" of launchers, SIRTF and other missions using expendable rockets were well isolated from the manned program. While the shuttle launches were put on hold, we continued to prepare for the launch of SIRTF. Because Gallagher and Irace had not been entirely happy with our practice review at the end of January, they staged a walkthrough of the revised materials at the management team meeting on February 6 and held another partial rehearsal the following week.

"I want to note that in the display room next door we are going to have a reception after the successful preship review, so we'll have to confer around five o'clock to see if we should proceed," John Straetker said to Frank Carr, the chair of the observatory preship review panel, on February 19. Not only were we planning a party this time, we were so confident that Straetker was teasing the review board about it. We were crowded into a conference room at Lockheed decorated with colorful posters from earlier space missions. The usual horseshoe of tables at the front of the room surrounded the projector. However, there were so many reviewers that only about half could sit at the front table; the rest were crowded with the rest of us into the sea of spectator chairs.

Counting the review of November 21, we had been through six rehearsals. Gallagher and Irace had wanted us to be much harder on ourselves than we expected a review panel to be. The approach had been polished and buffed to the highest degree of perfection we could muster. Unlike the November review, when the package had been pulled together at the last minute, we had submitted our material three times, each time revising it to make improvements.

Lockheed had decided to beef up security (not connected with our review), and latecomers held up by the security measures continued to trickle in after we started. Otherwise, everything went smoothly. A reviewer told me later that she could tell we were ready to launch from the first viewgraph. We approached the podium with an air of confidence. Irace and Straetker were relaxed and self-assured in starting us off, and each speaker addressed questions with aplomb. Because we had gone through the material together, we also knew which questions to deflect to later presenters or to direct to others in the room.

The review was organized around the findings from the previous one, and Straetker presented a detailed list that mapped each issue into the appropriate place in the thick package of viewgraph copies. Because it had been by far the largest concern, we started the in-depth presentations with our flight software. Lockheed systems engineer Nick Vadlamudi had put together a particularly effective summary viewgraph, which he now put up in the form of a field of questions: Did we build the software right? Did we build the right software? Is the flight software robust? Is the flight software ready? Under each key question were three or four subquestions, all of which he promised would be answered in the following presentations.

And so they were. Most of the discussion centered on minor items that had little potential for mischief. The two threatening bugs found by Kasuda's group were left in place, now with their implications more fully understood. Tests the previous day had indicated that the problem with the heap was not likely to arise in practice. With regard to the second problem, it had been determined that if the onboard memory got nearly full and the system rebooted, it went catatonic while it tried to salvage the data that had been put into memory prior to the reboot. There was a possibility that this state would last long enough to cause a new reboot, putting the system into an infinite loop. However, we could avoid this situation by being careful to avoid filling the memory. A software patch could be put in place after launch to be sure we never entered the danger zone.

Gallagher summarized: "A week ago we thought these were potential showstopper problems. Now we have a solution to one of them, and the second may not be an issue. We have a lot of people looking at these very hard, and they are making a lot of progress." Irace inserted: "We are definitely pre-loaded to using the software as it is and not making a change before launch." Gallagher continued: "We are virtually unanimous that these are not constraints to ship, and . . . that they are not constraints to launch."

The only significant remaining issue was the paperwork. A lot of progress had been made, but a lot more was needed. The intent of the paperwork is to give everyone involved an opportunity for orderly reflection on any departures from the original design expectations. In practice, however, the material is usually not processed in the timely fashion that might allow such reflection; paper accumulates in untidy heaps in people's offices. Finally, when the paper is the primary obstacle to launch, it must be processed very quickly just to get it out of the way.

At the end of the review the board made the traditional round of comments. They all felt comfortable with our technical readiness for launch. The chair of the independent review team summarized: "The concerns about the flight software are removed: release 6.2.1 shows good stability; the fault protection and stress testing were successful; and the Kasuda review adds a lot of confidence. The latch valve has been requalified. There has been lots of progress on paperwork and plans are in place to close all the critical paperwork prior to launch. The launch slip has resolved concerns about operational readiness. The launch delay provided the needed relief to allow orderly completion of all open issues. The project took excellent advantage of the extra time."

Straetker closed the proceedings by inviting everyone to the reception, adding with a broad grin: "Thanks to the review board that gave us so many insightful comments last November that we were able to do much better this time."

The review board's report, dated February 26, 2003, concluded "that the established success criteria have been met and that the SIRTF Observatory is ready to ship to KSC for launch site processing. . . . The Project has made excellent use of the additional time resulting from the Pad 17B scheduling conflict and has achieved a state of readiness considerably better than many projects at this same juncture." It was official: the observatory was complete!

23 LAUNCHING PROVES DIFFICULT

■ On March 3, 2003, a tiny caravan left Lockheed Sunnyvale and blended into traffic. In the lead was a nondescript truck carrying critical ground support equipment. Following it was a new Penske rental truck carrying SIRTF itself. In the rear was a recreational vehicle. Various sensors attached to SIRTF, such as accelerometers to indicate how smoothly it was riding, sent their data to monitoring equipment in the recreational vehicle via short-range radio. The trip was uneventful—in fact, unnoticed—and everything was delivered to the Kennedy Space Center at Cape Canaveral three days later, announced by a short e-mail from Mike Werner: "Hi, all—SIRTF is in Florida!" The launch was scheduled to occur in just six weeks, on April 18.

The *Columbia* disaster continued to dominate the news. Initially, the stories were optimistic. CNN.com, for example, reported on February 17 that NASA had strong support:

> The U.S. space program, under fresh scrutiny in the wake of the *Columbia* disaster, enjoys bipartisan support on Capitol Hill, backing that appears unlikely to diminish significantly even as some lawmakers raise questions about the safety of manned space exploration. Amid relatively flat budgets for NASA in recent years, the support is sometimes

> more rhetorical than financial, but with key facilities in big states that are home to powerful lawmakers and an almost mystical appeal to both the political and national psyche, NASA appears poised to weather this latest crisis.

But the media reports quickly became more probing and critical. On February 28, CNN.com reported that

> NASA Administrator Sean O'Keefe endured searing criticism in a congressional hearing Thursday, the day after release of NASA memos that raised safety concerns with chilling premonitions of the space shuttle *Columbia* disaster. In the internal e-mails, shuttle engineers raised fears that the left wing of the orbiter could burn off with the loss of the crew. The memos, written in the days before the *Columbia* broke apart, never reached top NASA management. The orbiter was lost during atmospheric re-entry February 1 shortly after experiencing problems with its left wing. "I read this stuff before you did. That's crazy," said Rep. Anthony Weiner, D–New York, referring to reports in the press late Wednesday. Weiner heatedly demanded to know why the concerns had not reached O'Keefe while the shuttle was in flight. "Have you fired anyone for not bringing them to your attention sooner? I can't think of anything more important on your desk than how's that shuttle doing," he said. O'Keefe responded that the appropriate experts had considered all possible problems and decided that there was no landing risk. "We encourage, expect, demand that people exchange ideas and solutions on how to deal with anomalies that occur on flight," O'Keefe told the House committee hearing.

Meanwhile, we set up and confirmed that everything had survived the trip. The project began to pick its way adroitly through the forest of prelaunch reviews. A prelaunch press conference was held at NASA Headquarters, but hardly any reporters attended; we were upstaged by the tensions in Iraq. The launch itself was to be dedicated to Earle Huckins, whose distinguished career at NASA had included a crucial intervention to solve the budget problem that had threatened to block our getting under way in 1996.

Nine days before launch we ran into our first real problems. Gallagher described them in a teleconference:

> Up until twenty-four hours ago everything was going beautifully, and even now everything is going beautifully technically. We stood down for one day due to a threat of lightning; there was also a problem with the ITT Cannon launch [electrical] connectors, but that was resolved. So, SIRTF has been unbagged and the fairing is being installed. The Flight Readiness Review is scheduled for tomorrow at 10 A.M. Yesterday, it became clear that NASA Headquarters is undergoing a huge amount of anxiety. The observatory is perceived as being so clean that they are concentrating on the launch vehicle. I'm not acquainted that closely with launch vehicles, but I'm impressed with the job Boeing and the center crew have done. However, there was a delta [Integrated Mission Assurance Review] yesterday. We knew something was wrong because many high-level headquarters people were participating. I believe it's quite likely that the outcome of today's meeting is that the flight readiness review will be postponed and that it will be impossible for us to make our launch date.

Confirmation came in an e-mail from Gallagher to the SIRTF team the next day:

> I regret to inform you that the SIRTF Launch has been delayed until the April 26/27 time frame. NASA HQ has raised some questions about the graphite epoxy motors on the launch vehicle and has requested the delay to allow time for these issues to be addressed. . . . The good news is that the Observatory is ready in every way. The fairing was successfully installed today and we are safely inside awaiting our ride. I am certain that all of you are as disappointed as I am, but we must keep focused on being prepared to support a successful launch, In-Orbit Checkout, and SIRTF mission.

We heard that there was an atmosphere of fear at NASA Headquarters caused by the realization that any failures following *Columbia* could turn the public against NASA, just as had occurred with other failures in the early 1990s. Attention was concentrated on the graphite epoxy motors, where a modest level of delamination had been discovered in exit cone liners on the nozzles.

Delta II rockets are among the most reliable in the world. The failures that have occurred have been in the electrical or hydraulic systems. Only once has a graphite epoxy motor failed, and in that instance its case split—an event unconnected to the issue with our boosters. In fact, so far as we could tell, nozzle delaminations were a way of life with all solid rockets, not just those used with Deltas. The causes had never been identified, so no cure had been found. Nonetheless, they did not seem to pose much of a threat; it was thought that they sealed up when the nozzles heated during launch. Later, we found that 30 percent of the solid boosters stored for use with Deltas had similar delamination.

Gallagher's next e-mail message, dated April 18, brought crushing news:

> It is with great regret I am writing to let all of you know that we will be standing down from our current launch opportunity. It was decided today by NASA HQ that the risk related to delamination in the Graphite Epoxy Motors is significant enough to cause the SIRTF launch to be delayed. I am sure you are all as profoundly disappointed as I am and I want all of you to know that this is no reflection on the quality of the Observatory or the readiness of the entire Operations team. The decision is that we will launch in August following MER-B. As more information becomes available, I will update you. I thank all of you for your continued hard work.

Two of the nine solid booster motors that had noticeable delamination would be replaced. Our logo was to be stripped off the rocket, and the Mars Exploration Rover (MER) logo substituted. The penalty for a MER missing its launch window was huge—the next suitable opportunity was in four years. That project now had an overriding priority relative to an astronomy mission that could be launched on nearly any date on the calendar.

It was a stunning setback. The true reasons may have become apparent in a *USA Today* article dated April 24, 2003. Headlined "NASA Took Too Great a Risk, Witnesses Say," the article went on to report that "Robert Thompson, who ran the shuttle program until 1981, said he could not imagine allowing the shuttle to fly knowing that a briefcase-sized hunk of foam could strike the wing at nearly 500 mph. . . . Thompson likened NASA's attitude about foam damage to that of a person who narrowly escapes several gunshots and then assumes future gunshots pose no danger. 'Would you like me to continue to shoot at you?' Thompson said." It is hard to avoid a suspicion that NASA Headquarters had been forewarned about this testimony. Even with a minuscule chance of a launch disaster, there was enough e-mail traffic on the delaminations to fuel a dandy investigation should a failure occur.

We had a short discussion about the possibility of using the delay to fix every annoyance we knew about in the SIRTF flight system. However, Gallagher and Irace settled on our doing as little as possible, on the principle that "better is the enemy of good enough." We would do the software patches scheduled to be put in on orbit to prevent the problems with our mass memory card. Otherwise, we would continue to run tests, upgrade the ground system, and get vacationing out of our systems.

But there were other problems. The "August" launch did not mean we had a guaranteed slot on the manifest. In fact, to make it a reality, both Mars rovers, MER-A and MER-B, and another GPS satellite had to launch. Furthermore, a wiring problem had just been discovered in the MERs, removing any optimism we might have about their getting off early. It looked like we might just make it to the pad in early September—and our window closed on September 9. It was not even clear that they would put us on the pad under those circumstances, particularly since the dates were near the peak of hurricane season, when weather delays are common. If we slipped to our next opportunity in late October, however, we would be launching into the center of a huge traffic jam to use the Deep Space Network of radio antennae. The conflicts would threaten our in-orbit checkout, and perhaps even the ability to monitor our spacecraft for safety. Given the unfortunate flexibility we had in terms of launch dates, it began to look like we should start planning for February.

> News release: "Russia to Launch SIRTF," by Leopold Strauss,
> *Space Flight Gossip*
> Baikonur, Kazakhstan, 30 April 2003
> In a move that is sure to send more than ripples throughout the NASA community, the Space Infrared Telescope Facility (SIRTF) project, managed from NASA's Jet Propulsion Laboratory, has signed a contract with the Russian Space Agency to launch SIRTF on board an R-7 Proton rocket. Side-stepping the Kennedy Space Center and Boeing, the prime contractor for the U.S. Delta II launch vehicle, SIRTF project manager David Gallagher signed a binding contract with Gen. Alexander Medvedev, director of the Russian Baikonur Cosmodrome launch complex, to launch SIRTF into an Earth-trailing orbit on May 15th, 2003. It is reported that the Russian launch will cost only $1–2 million,

> a savings of over $30 million as compared to the Kennedy Space Center and Boeing cost.
>
> Recent launch delays associated with cracks in the Boeing-made Delta-II solid rocket booster motors have frustrated project managers, and apparently led Gallagher to contact Gen. Medvedev about a possible Russian launch. The May 15th launch will come at least three months before the earliest scheduled Kennedy Space Center launch window, saving the project an additional $25 million in contractor salary costs associated with the delay. "Since the SIRTF telescope is ready to go, and the Russians are slightly less risk-averse than most of the officials at NASA Headquarters, it was the natural thing to do," claimed Gallagher in a phone interview from his office at JPL. "After all, isn't better, faster and cheaper, what we are all about?" He added, "Is there anyone out there that can honestly say they want to be in Florida in August?"
>
> When asked to comment on the upcoming Russian launch, Medvedev responded, "At last we are getting back into the space science business. Are we truly supposed to simply be some sort of space taxi for the space station until the shuttle is fixed?"

Gallagher told us he had first read this spoof on a bulletin board and had thought to himself: "You're really in trouble now, Dave." At least it gave us something to smile about in an otherwise grim time.

To avoid a year's delay, we had to question the two boundary conditions that were pushing us in that direction: the launch manifest and our launch window. Although the MERs had the highest priority, the status of the GPS was less clear, and we could legitimately claim that all of our problems stemmed ultimately from the air force's refusal to yield the launch pad in late January. We urged the Kennedy Space Center launch director, Omar Baez, to get the GPS launch postponed. With the nominal timeline it did not conflict with our needs; however, it was to be supported by the same Boeing launch team. Having to go at top speed for the two MER launches, the GPS one, and then ours (plus the two previous GPS launches and our staging) was going to sap their reserves, leading to inevitable delays. By late May this negotiation had succeeded. It was a concrete indication that our "August" launch was being taken seriously.

We now got to sit on the sidelines and learn more about the difficulties of getting into orbit by watching MER-B. On June 21, a routine inspection revealed that some of the cork insulation on the main body of the MER-B rocket was not glued on firmly enough. This problem was repaired in time for a June 28 launch, but a boat wandered into the security zone, and by the time the zone was clear the upper atmospheric winds were too strong. The next day, more problems were discovered with the cork insulation. To allow time for repair and testing, the launch date crept backward, settling on July 6. However, on July 5 it was discovered that a battery had discharged in the system that is supposed to blow up the rocket if the launch goes poorly. Further emergency repairs were also required to the insulation. That put the launch back to July 7, when MER-B finally left the launch pad.

With MER-B launched and the GPS satellite postponed, we had a firm date of August 23. A series of mishaps (lack of a suitable ship, repairs to the one that was found, bad weather) delayed the arrival on station in the Indian Ocean of a downrange telemetry receiver, causing the launch to be postponed to August 25, one week before the traditional peak of the hurricane season (Bossak 2003).

We had invested decades in SIRTF, and its launch would be one of the key events in our lives. We all wanted to be there. The SIRTF team and guests assembled in Cocoa Beach a week before the launch and entertained each other with a blur of receptions, press conferences, tourist jaunts, and parties. Cocoa Beach is a typical beach resort with huge hotels crowding the ocean and tacky shopping malls and cheap restaurants lining U.S. Highway 1 just inland; only occasional business names and astronaut pictures indicate the area's unique attachment to the space program. The Kennedy Space Center is just a few miles north of town, if one knows the shortcut past the cruise ship wharves and through the naval base. The launch pads and support buildings are scattered about a vast mangrove swamp patrolled by alligators and festooned with long-plumed herons and egrets. The salt air rapidly dissolves the center's monuments, and the grounds are decorated with rusting hulks of historic launch pads.

As the launch date approached, the air over the Florida coast became seriously unsettled. But then we entered into an improbable stretch of good luck. Unusually well-behaved weather settled in around the twentieth. There was a slight exception in the form of a dramatic lightning storm that swept onto the Space Center from the west and north on the twenty-fourth. Our pad was at the south end of the complex and was the only one not closed. Even the weather seemed to be aware of our twenty-year wait and was trying to speed our departure.

And then, three days before the launch, we got more bad news. A shake test of an electronics box for the rocket for the Gravity Probe B mission had revealed an intermittent electronic fault. A small electrical component had come loose from a power supply and had rattled around, briefly shorting out various electrical signals. The power supplies in our rocket electronics had been made in the same plant and at about the same time. Gallagher called us together: "We are green on everything there is to be green on—the weather, the range . . . with one big exception that just dropped into the punch bowl yesterday. There is a concern that the surface-mounted capacitors [in the rocket electronics] may not be bonded down properly. Twenty of our parts have been implicated by date lot code. Eleven of the parts have been cleared by checking X-rays. Nine of the X-rays can't be located. I can't add much value; I have to leave it to the experts. I suspect it is just 10–30 percent they'll get through all the hoops in time." Before this event, Gallagher's upbeat demeanor had been one of his most powerful leadership tools. Now he seemed morose except when the limelight played directly on him. Our efforts had reached a crescendo as we neared the launch, and it was going to be nearly impossible to hold the team together for a similar push if we had a major delay. Black humor circulated: Boeing had made a banner to celebrate our launch as the 300th of a Delta, but we claimed that to get a quantity discount they had bought a banner for the 301st launch just in case ours failed.

We got together again on a late-night telephone call. Gallagher reported that "Boeing came in on this very impressively. They have fifteen people working overnight. Detailed analysis of 1040 X-rays is convincing everyone that the original approach to screening was OK. They have a very robust system—they really have their act together. It's like a management plan that in the middle has a balloon that says 'and then there is a miracle.' It looks like the miracle might happen. Keep praying to help it along."

At five o'clock the next afternoon, about eighty people assembled in the Mission Briefing Room for a Delta flight readiness review, the final review of our launch status. Big projection screens for weather and launch status information dominate this huge chamber; beneath them runs a long conference table. Launch Director Omar Baez ran the meeting from one end of this table. The purpose of a flight readiness review is to bring together all the far-flung elements of a mission and make sure everyone agrees that it is ready to go. First came a combination of reports that indicated that the rocket electronics had been judged acceptable. Baez took a roll call of about twenty people sitting at the table and representing key disciplines; all indicated they thought the launch could occur. Gallagher followed by polling half a dozen high-level project and JPL management personnel, who agreed.

The launch had to be at 1:39 A.M. to get into the right orbit. It would be controlled from an unpretentious galvanized steel–sided building in a complex perched on one of the islands of dry land within the Space Center. I had gotten myself assigned to a seat at a control console. As launch time approached, I found myself seated with a number of others who had played important roles in building the observatory and had also been rewarded with an up-close view of the launch process. All of us took great care to be briefed in detail on how *not* to have any influence on events, then settled in with headsets (with deactivated microphones) to listen. Each of us had a monitor and grid of switches underneath to control what we saw and heard.

The progress was amazingly easy to follow. By activating the appropriate switch we could listen to any of the main participants over the communications network, and by activating many switches we could listen to all or any subset of them. They all followed a set of rules (that were laid out in a fat book at each console). As a result, it was easy to follow virtually any combination of communication inputs—statements were short and to the point, and the person addressed was always identified by role in the launch rather than by name. A large screen at the front of the room switched sequentially through weather maps and various views of the rocket.

Everything went perfectly as the countdown continued. Twenty minutes before the launch, about ten of us spectators piled into a couple of rental cars and drove across the swampland to find a good place to watch. Our objective was easy to spot—the launch pad was lit with searchlights. We came to a barrier across the road, parked, and piled out. A Space Center police car had followed us, and the officer suggested we move farther away from the launch. We demurred. "Well, if an anomaly occurs, get out of here quick!"

Three brilliant searchlight beams from the launch pad pierced the night sky, which was studded with stars peaking from behind a few slow-moving clouds. SIRTF and its rocket had been transformed into a glowing obelisk. The air was still and thickly humid, with a foreboding of rain. In the distance, a female voice was counting down over a barely audible loudspeaker. For my companions, a mix of engineers and managers who had helped assemble the observatory and tend it at the Cape, the launch was the climactic end of their work. They wished me luck in using the observatory, a gesture I took to be symbolic of the transfer about to occur from the engineers to the science community. In my symbolic role, I could thank them each for getting us to this point. We lined up against the barrier and held our collective breath as a hush seemed to descend over the entire Cape. We tuned our ears intently to the distant counting and our eyes to the obelisk on the pad.

As the count hit zero, a bright flash of flame and a cloud of vapor emerged from the back of the rocket. The flame grew, intensely lighting the whole sky as well as our surroundings. It was followed quickly by a loud roar as the rocket lifted slowly, then more rapidly. As it picked up speed, it diminished in our view to just that bright flame and its roar faded to a sound like the *blat* of a small automobile with a faulty muffler. As the rocket headed east and away from us, we could see it jettison six of the solid boosters, whose hot nozzles glowed as they tumbled down toward the ocean. The passage of time was suspended; after a two-minute, forty-second eternity, and close to the limit of vision, the remaining three boosters were dumped. The swamp slowly settled back to its thick, humid hush. As we turned to go back to our cars I noticed a sign in the bush I had been brushing against: Beware of Alligator.

We nibbled nervously on snacks that had been laid out at the entrance to the launch control building as we watched a TV monitor suspended in a corner. Jim Houck appeared, looking disheveled in a tattered, too-tight T-shirt, which I suddenly realized he had preserved from the IRAS launch twenty years ago. Everything went exactly on plan for forty minutes. Then the Deep Space Network stations in Australia failed to pick up the spacecraft telemetry. The announcements accompanying the monitor display somehow increased the tension by turning their focus to the spent rocket, which was being directed into the Pacific Ocean. Although they were just reporting what news was available, it felt like the payload had been abandoned. We all twisted into knots inside. "I have no nails left," Gallagher later reported. "We had a little extra delay there acquiring from Canberra" (Ray 2003). The problems turned out to be with the Deep Space Network station (an error had been made in the units of the pointing coordinates). Our satellite was headed on a perfect trajectory into its desired orbit.

24

AUGUST–DECEMBER 2003: CHECKING THINGS OUT

■ A few days later back at JPL, we were focused on two "mission-critical events." We had noticed years before that a disproportionate share of the failures in previous missions had been in events required to get the satellite properly deployed—things that were done only once, and often that were difficult to test thoroughly on the ground. Examples include the WIRE premature cover deployment, the Mars Climate Orbiter insertion into the wrong orbit, and the Mars Polar Lander crash landing. We had therefore put extra effort into planning and reviewing our own events of this type: the launch itself, opening a couple of valves to keep the dewar from overpressurizing, getting the solar cells pointed to the sun, establishing radio contact with the Deep Space Network, ejecting the cover over the front of the telescope, opening the aperture door that let light reach the instruments, and focusing the telescope. The first four had already come off perfectly to get us into orbit, and we now ticked off two more: ejection of the telescope dust cover and opening the aperture door.

With these events behind us, we had a functional observatory that had to be checked for correct operation and characterized and calibrated before it could be used for science. Because the telescope had been launched warm, however, its infrared emission still blinded our detectors. We impatiently waited for it to cool enough to get started. Finally, IRAC was able to "see" the sky and took an image that was stupendously good given that we had made no adjustments. NASA administrator Sean O'Keefe, who was still being grilled about agency shortcomings connected with the *Columbia* disaster, commandeered this image for immediate release as a desperately needed scrap of good news.

Plans for our in-orbit checkout (IOC) had been laid out in great detail prior to launch. We followed this timeline, or rather we followed the revised timeline that was laid out early each morning and then confirmed in a meeting before noon. Each set of commands had to be run through a laboratory at Lockheed Denver, where there was a duplicate of the spacecraft electronics and software, before it could be submitted for transmission to the satellite. Then there was a "command conference" in which a group (some in a conference room at JPL and some in Denver connected via a telephone link) walked through the command sequence. Finally, the new set of commands was sent out to the Deep Space Network for transmission to SIRTF. The process was like a waltz by highly trained elephants. It was slow, but it was exquisitely choreographed and beautiful to watch.

Although deviations from the master plan were continuous, overall our IOC went incredibly smoothly. Bob Wilson led the operations team in his unique manner: firm, but at the same time almost excessively polite and considerate. Sue Linick oversaw the mission timeline; her style was more frenetic, exemplified by the constant supply of sugar (in the form of cookies and candy) she fed her team to keep them on a steady high. Prior to launch, Wilson and Linick had made it clear to us that their procedures would be rigidly controlled, with long review cycles before any new command sequences could be accepted. In practice, their teams scrambled to implement whatever was needed to recover from failed observations and adapt to current circumstances. The "inflexible" IOC timeline was overhauled approximately every other day.

As soon as the telescope had cooled sufficiently to allow all three instruments to operate, we analyzed our first images to see if we could agree on how to move the secondary mirror to put the telescope into focus. This adjustment was our final mission-critical event; if something went wrong, we could drive the telescope *out* of focus instead. We were already close enough that the IRS and MIPS instruments had little to gain, but the IRAC images were degraded. Just prior to our "focus summit" meeting, I consulted a fortune cookie for advice: "You will soon make a move for the better." And we did.

Our overall success was reported at yet another review on October 23. Mike Werner summarized the situation to the board: "We have a beautiful, powerful observatory that is capable of doing wonderful science. I'd like to thank the engineers, mission ops team, and everyone else for providing it for the scientists to use." Even the problems that had arisen were in some sense planned. "Believe it or not," Gallagher said, "the anomalies we have encountered are ones we had written contingency plans for. . . . But to prove we aren't dry-labbing this, we had a suspend last night." And indeed that appeared to be the nastiest aspect of our machine: it always seemed to know when we were going to have a review, and to respond with some minor hiccup. Nonetheless, under the principle that every review is a CDR, the proceedings droned on, probing every minor anomaly.

Each of the instruments had entered orbit with a few problems: the shutter in IRAC plus reduced throughput in two of its channels, and the filter in IRS and difficulties in achieving an optimum calibration for the data. Although still capable of carrying out the science it had been designed for, my instrument arrived on orbit with a relatively large number of such issues. Another cable had failed, taking out half of our 70-micron array. The cosmic rays on orbit had a much more dramatic effect on our far infrared detectors than had been indicated by our prelaunch testing, and they were about three times less sensitive than we had hoped. The first issue I found distasteful; a preventable problem had compromised a huge amount of work on our part. Although the second issue had substantial implications for many science programs, it was not personally disturbing: I saw no way that my team could have done any better than we had.

There was also a third issue: a design flaw in the optical train allowed short-wavelength light to leak around the filter in our 160-micron channel and reach the detector, at least when the instrument was focused on very bright stars. Although this portion of the optics had been studied and reviewed very carefully, once the problem became apparent on orbit the cause was glaringly obvious. Fortunately, the problem was likely to affect only a very limited set of programs because the leak was relatively small. Nonetheless, I was deeply depressed because we had missed perfection. For a while early retirement appeared very attractive—as a form of professional suicide, I suppose. At the same time, I could see from a detached perspective that we were in pretty good shape to do groundbreaking science. We had not blown up on the launch pad, the satellite was working beautifully, and the instrument had all of its basic capabilities intact. I had to keep visibly upbeat so my team would not sense my distress and stop working at peak efficiency. My distress was relieved when, in a dozen hours of observation during instrument checkout, we made as many new discoveries. I finally realized that I wanted to be a part of any breakthroughs we achieved.

Our progress through checkout was, of course, marked by a series of reviews. So far as my team was concerned, the most important one marked turning responsibility for "our" instrument over to Bill Latter, the head of the instrument support team at the SSC in Pasadena. This group had thoroughly meshed with the instrument team in Arizona, so no real change was going to occur; however, a review seems to be the only way NASA can recognize a milestone even if it happened long ago.

In mid-December we had a science strategy review. The panel of astronomers was ecstatic about the rapid progress toward commissioning SIRTF and was greatly impressed by the small tastes of its capabilities and early observations we teased them with during lunch. Finally, this review had the atmosphere of a real inquiry into our status and a resulting celebration, not a CDR.

25
OUR COMING-OUT PARTY

■ The coming-out party for our new mission was on December 18, 2003. We gathered at NASA Headquarters for the official unveiling of a new name for the observatory and the release of its first data.

Two years earlier there had been a contest to replace our well-worn SIRTF acronym. There were more than seven thousand entries, each accompanied by a short essay on the virtues of the suggestion. A panel of educators, authors, and project members had narrowed this list to six, which were forwarded to NASA Headquarters for a final choice. The winning name had been sent in from the Northwest Territories by Jay Stidolph, who wrote:

> My suggestion for the renaming of SIRTF is the Spitzer Deep Space Observatory, named for Dr. Lyman Spitzer. Dr. Spitzer is, I believe, the father of the modern space-based telescope. Dr. Spitzer's revolutionary paper, written in the 1940s, was the first to propose the idea of putting telescopes into space, and thus above the blurring effects of the Earth's atmosphere, which not only revolutionized the science of astronomy, but it also pulled back the atmospherically induced blinders we had lived with for so long and revealed the true wonder and beauty of the universe. The hard-won science and images of breathtaking beauty that

> have been garnered from the Hubble Telescope have helped bring millions of supporters to the space program and taught us things about the universe, which we might never have discovered without it. (Spitzer Science Center 2003)

Spitzer himself best described the origins of his interest in space astronomy: "As World War II drew to its end, I was approached by a friend on the staff of the RAND corporation, an Air Force 'think tank.' He told me that his group was carrying out a secret study of a possible large artificial satellite, to circle the earth a few hundred miles up. 'Would you be interested,' he asked me, 'in writing a chapter on how such a satellite might be useful in astronomy?' With my long and ardent background in science fiction, I found this invitation an exciting one and accepted with great enthusiasm" (Spitzer 1989).

The naming ceremony led into a press conference releasing the first science results, which was followed by a reception. It was an afternoon of quintessential good feelings. The front rows of the auditorium were filled with three generations of Spitzers (including his widow, Doreen) and their friends. The family distributed big souvenir campaign pins bearing quotes from his space astronomy paper. Spitzer had had close ties with Ed Weiler, who had worked for him as a postdoctoral fellow, and John Bahcall, who as a more junior faculty member at Princeton had joined enthusiastically with him in promoting the space telescope. In addition, his modesty and politeness had endeared him to everyone who worked with him. Thus, the speeches were filled with warm reminiscences.

The first results from the telescope were visually appealing and scientifically intriguing. They were the product of an elaborate chain of steps to select targets that would interest the public. The most striking of all was described in the press release: "In a dark, elongated globule called the Elephant's Trunk Nebula, the new Spitzer observations reveal a glowing stellar nursery that was previously hidden from view by dust. The infrared view of the dark globule, which is located 2,450 light-years away in a nebula called IC 1396, shows previously unseen young stars and protostars and turns the dark cloud into a bright one that resembles a flying dragon." We also demonstrated how M81, a spectacular nearby spiral galaxy, transforms its appearance like a giant space chameleon as we observe it across the infrared spectrum. We showed that the material surrounding a nearby young star and that in a galaxy so distant that we see it as it was more than five billion years ago contain similar molecules.

This result previews use of the observatory to study the formation of molecules in the Universe at times that preceded—and hence set the stage for—the formation of the complex molecules of life on Earth. We had looked at a nearby star known to have a torus of material orbiting it and had found that the torus is filled by a disk of warm dust similar to the zodiacal dust in the solar system and suggestive of planets that might be deflecting comets inward.

These results were based on only a few hours of observing time. Spitzer, in a five-year mission, had some forty thousand hours yet to go.

26
OUR FIRST YEAR IN ORBIT

■ I bring this book to a close at the end of 2004, with a summary of Spitzer's performance to date.[1] The observatory itself is working almost exactly as we had hoped. The pointing, image quality, data handling, and all the other aspects of the spacecraft and telescope are close to the prelaunch predictions. Hiccups leading the observatory to protect itself by going into a "suspend" or "safe" mode—states in which observations are stopped—have been rare. Most important, a measurement of the liquid helium level in October 2004 showed it to be adequate for four more years—indicating that the full five-year science mission remains a possibility.

Of course, the observatory's performance must eventually be judged by the science it makes possible. Although a year is much too early for scientists to sort through all of the implications of new observations far beyond the reach of previous ones, we can at least make a preliminary assessment. To do so, let us again look at the four science programs we adopted to define the mission in 1993. Such a review will also illustrate one of the thrilling aspects of front-line science—its unpredictability. Even though the mission was designed around these programs, our contributions to them so far have been uneven.

Brown Dwarfs

The great successes of ground-based astronomy in identifying and studying substellar brown dwarfs since 1993 made it substantially more difficult for Spitzer to make major new contributions. By the time of our launch, the goal was no longer to demonstrate that these objects exist, by detecting significant numbers in the 200–800 times Jupiter mass range, but to find them below 10 Jupiter masses. The on-orbit sensitivity of Spitzer is sufficient for this goal. It was hoped that we could identify candidate objects efficiently by using IRAC to image nearby regions where stars are forming and the brown dwarfs are young enough still to be reasonably bright. We would winnow out the brown dwarfs by looking at the relative brightness of objects in the IRAC bands, their "colors."

Our initial efforts in this endeavor ran into an unexpected problem. Spitzer's sensitivity is sufficient to detect thousands of background galaxies projected onto every nearby star-forming region. Because they come with a variety of intrinsic properties and at a variety of redshifts, the colors of a significant number of these galaxies can mimic those of brown dwarfs almost perfectly. Astronomers are adept at solving such problems. For example, images made with large, sensitive ground-based telescopes in the near infrared, 1–2.5 microns, may provide enough information to reject the interlopers. If not, we may be able to identify the brown dwarfs by using deep X-ray images that pick up the emission of their active outer layers. As I write, however, we are not making the rapid progress we had hoped for.

The first year did yield measurements of older, more massive brown dwarfs that verified many of the predictions of theoretical models for their interiors and for the transfer of energy through their outer layers where it is radiated into space. Modest discrepancies remain that theorists can attack with the confidence that the foundations are correctly in place. This process is "quiet science" that can be just as important for further progress as the high-profile discoveries reported in press conferences.

Protoplanetary and Debris Disks

The Spitzer results on planet-forming disks have already been phenomenal. They have documented the very early clearing of the inner zones of some disks, shown a large variety at subsequent stages in the amount of circumstellar material, and even identified a few stars that appear to be the sites of recent huge collisions between planetesimals leading to the generation of large clouds of dust.

Figure 26.1 shows measured stellar system spectra (right panel) compared with the conceptual behavior (left panel) we had predicted. All the observed systems have excess emission well above that of the star by itself. The broad peak near 10 microns is a signature of small grains of minerals with silicate material in them (silicates, a compound of silicon and oxygen, are common in rocks on the earth's surface).

The laws of radiation dictate that an object emits energy at a wavelength inversely proportional to its temperature. That is, as a body is heated, the peak of its emission moves from the far infrared to mid-infrared to near infrared to optical to ultraviolet. The star shown at the bottom left of the figure is so hot that the peak of its output is at wavelengths off the spectrum shown at the bottom of the right-hand panel. All that is apparent is the drop in emission into the infrared. If the star is heating a disk of material, then the warm material close in emits at relatively short infrared wavelengths, while the cooler material farther out emits at longer wavelengths. The top four system spectra

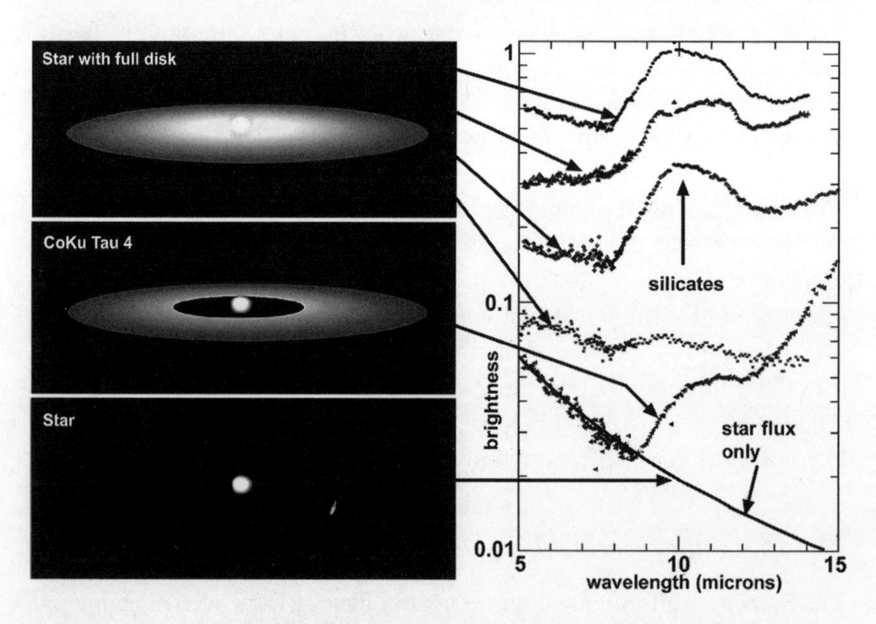

Figure 26.1. IRS Spectra of a Number of Young Nearby Stars. The line labeled "star flux only" is the output of the star alone. All the stars have shallower slopes than this curve, indicating excess emission due to circumstellar disks. (After a figure published on the Spitzer Science Center Web site, http://www.spitzer.caltech.edu, 2004; D. Watson)

at right show excess above their stellar outputs at all wavelengths; the trend of the spectra falls much more slowly toward longer wavelengths than the stellar-only spectrum (they are also offset vertically so that they do not overlap). They have warm dust both close in and farther out from the star, as shown by the star with full disk at upper left. Because the spectrum of CoKu Tau 4 (portrayed at middle left; spectrum shown fifth down at right) shows very little emission above the star flux out to 9 microns, its disk is largely cleared of material out to a distance nearly equivalent to that of the earth from the sun. A plausible explanation is that one or more planets are already forming close to the star, and as they circle it their gravitational fields throw the dust out of this zone.

Figure 26.2 expands this picture to a large number of stars over a span of millions of years. Rather than showing spectra—output versus wavelength—it takes a large sample of stars measured at 24 microns. The actual signals from the stars have been divided by the signals expected if the stars have no material around them. Thus, a star with no surrounding disk would come out at 1, within the errors in the measurements (stars that fall below the line

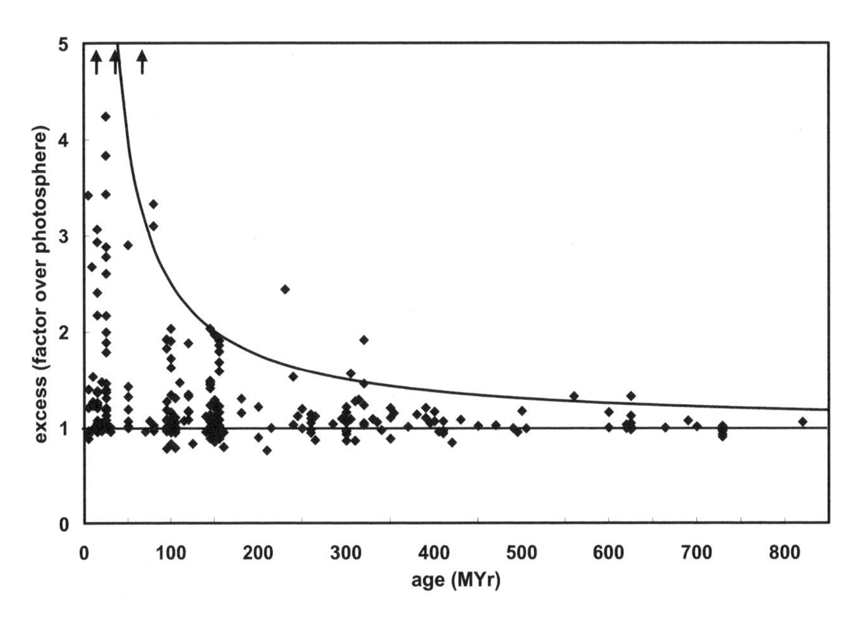

Figure 26.2. Pattern of Disk Excess Emission with Stellar Age. There is a general decline in the disk emission around these stars and a huge variety in the amount of disk emission at all the young stages. The line above the points shows decay with a characteristic time of about 150 million years. (Published on the Spitzer Science Center Web site, http://www.spitzer.caltech.edu, 2004; G. Rieke)

at 1 do so because of either measurement errors or errors in extrapolating the stellar output). Stars with disks have additional infrared emission above the pure stellar output at 1. The figure shows that many stars younger than about 100 million years have disks, but the incidence of large amounts of additional emission above the stellar output drops with age and the excess emission is modest after 150 million years.

Spitzer images at 70 microns of two nearby stars (both about twenty-five light-years distant) are shown in Figure 26.3. Fomalhaut (upper right) has a ring of dusty material around it at a distance from the star of about 110 times the distance of the earth from the sun. The ring is moderately filled in by dust grains falling into the star (this behavior is more obvious at 24 microns, where this dust dominates the image because its proximity to the star means that it

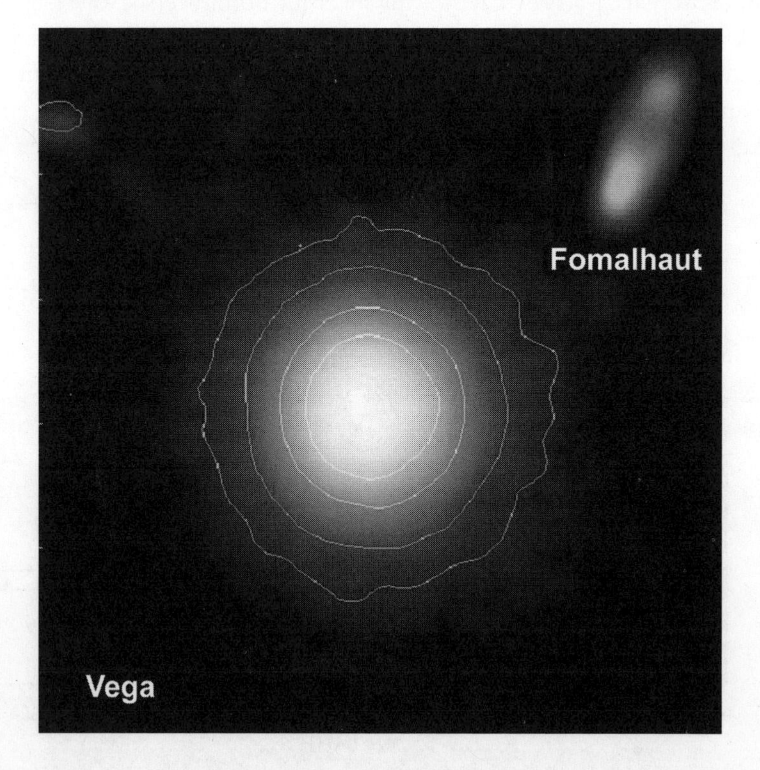

Figure 26.3. Comparison of Images of Vega and Fomalhaut at 70 Microns. The images are to the same scale; Vega's debris system extends from the star to nearly one thousand times the distance from the sun to the earth. (Published on the Spitzer Science Center Web site, http://www.spitzer.caltech.edu, 2004, 2005; K. Stapelfeldt, K. Su)

gets relatively warm and hence emits well at shorter wavelengths). Vega's disk (center) is far larger than the one around Fomalhaut. The detailed properties of the image (and other ones at shorter and longer wavelengths) indicate a ring of material orbiting the star about eighty times the distance of the earth from the sun that is basically similar to the ring of dust around Fomalhaut. However, a catastrophic collision between two large objects (perhaps 1,000 kilometers in diameter) also orbiting in this ring seems to have set up a series of collisions that are producing small grains in huge numbers. These grains are being thrown out into interstellar space by the stream of photons leaving the star.

Considered together, these results have a remarkable correspondence to theoretical models of the formation of the solar system that show a period of violent collisions for about 100 million years (the moon was formed in such an event when the system was about 40 million years old). Studying the debris disks around other nearby stars appears to offer a superb opportunity to test and expand theories for the formation of our own solar system.

Ultraluminous Galaxies and Active Galactic Nuclei

The power source for ultraluminous infrared galaxies and their relationship to active galactic nuclei has been under debate since the 1970s. Astronomers used spectra obtained with ISO to argue that these galaxies could be divided unambiguously into those dominated by the formation of huge numbers of massive stars ("starbursts") and those dominated by accretion onto supermassive black holes (active galaxy nuclei, or AGNs).

The Spitzer spectra reach ten to a hundred times deeper than those from ISO. In some cases they confirm conclusions reached previously with spectra of far lower weight. However, in other cases the new spectra indicate that the sources are more complex than previously thought. Some of the ultraluminous galaxies have absorption features due to ices and hydrocarbons—features that have previously been found mostly in star-forming molecular clouds in the Milky Way. Others have strong absorptions due to finely divided silicate dust grains, while others have spectra dominated by large molecules mostly of carbon and hydrogen. The more sensitive spectra also complicate rather than simplify the identification of active nuclei. Although some of the ultraluminous galaxies show emission lines that need to be excited by an AGN, not stars, some other galaxies known to have active nuclei from optical measurements do not show such lines.

Thus, for this area of investigation, the improvements brought by Spitzer have initially shown that answering the relevant questions is substantially more complex than had been suspected. This perverse form of progress will help focus programs for the rest of the mission that presumably will yield a more satisfactory picture.

Galaxy Formation in the Early Universe

The biggest surprise in studying the distant and early Universe has been how far and how early Spitzer can see. In fact, Spitzer has detected perhaps the most distant galaxy known, at a redshift of nearly 7. Otherwise, the mission has made contributions to this area roughly as expected, with a significant expansion of our knowledge but few major surprises.

The mission is documenting the kind of galaxy that dominates star formation at different ages of the Universe. Out to a bit beyond redshifts of 1, eight billion years ago, when the Universe was only about five billion years old, we are confirming that most star formation is occurring in normal spiral galaxies but at a greatly elevated rate over nearby spirals. This result can be interpreted as seeing these galaxies forming the disks and spiral arms that dominate their appearance in the familiar pictures taken in blue light, where we see mostly relatively young and massive stars. When the Universe was younger, approximately from two to five billion years old, massive galaxies similar to local ultraluminous infrared galaxies were in their heyday. They may correspond to the formation of elliptical galaxies and to the central parts of spirals.

We are also discovering at least some of the previously hypothetical ob-

Figure 26.4. Comparison of X-ray, Visible, and Infrared Images of a Distant Active Galaxy Nucleus. (Published on the Spitzer Science Center Web site, http://www.spitzer.caltech.edu, 2004; A. Koekemoer, M. Dickinson)

scured active galaxies. Figure 26.4 is an example. The Chandra X-ray image shows a distant active nucleus; the very deep Hubble image does not find a trace of it; but it is nicely detected with IRAC in the 3–5-micron range. We find that these objects are usually at redshifts typical of quasars. To avoid discovery by Hubble they must be in galaxies without young stars—that is, relatively old elliptical galaxies. In addition, their active nuclei must be so obscured that little light comes out in our direction.

Other Discoveries

Just as the unpredictability of science changed the impact of our four defining programs, it also caused us to stumble into discoveries we had not anticipated. One is illustrated in Figure 26.5. The figure shows two regions forming large numbers of young stars, with many objects organized in lines on the sky. This behavior suggests that at least some stars form along strings or sheets of interstellar gas. This possibility has been proposed in theories that suggest that magnetic fields can organize interstellar gas into filaments that fragment into stellar-sized clumps.

Normally, we would expect new stars to be scattered randomly because they are launched with speeds of 1–2 kilometers per second, which in a million years would carry them about five light-years from their sites of forma-

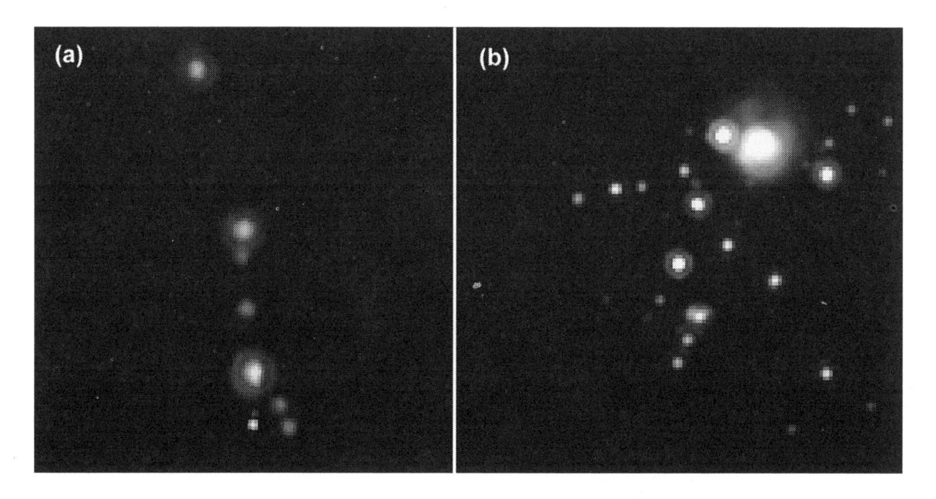

Figure 26.5. Images of Very Young Stars. (a) Wavelength of 70 microns, region of Herbig Haro Object 24. (b) Wavelength of 24 microns, region of NGC 2264. (E. Young, personal communication, 2004)

tion. At the distances of the regions in Figure 26.5, five light-years is nearly the width of the images shown. Thus, even if young stars start their lives somehow arrayed in lines, within 100,000 years these lines would start to disintegrate. Our images thus seem to show stars really close to the moment of their formation, perhaps no more than 50,000 years old.

The strings of forming stars are a discovery, but one anticipated by previous work. The purest discoveries in science happen by accident and are so far outside expectations that initially they defy explanation. We made one of these in the first months of the mission, although it took a year to appreciate its significance.

Cassiopeia A is the most recent known supernova explosion in our galaxy. Cas A lies in a region obscured by interstellar dust that hid it from visible observation, or it would be far more famous. The first astronomer royal, Flamsteed, possibly noted it on one occasion in 1680. It was rediscovered as a bright radio source, and the time of the explosion has been set by observing its expansion over the years and tracing all the fragments backward to a common origin. They come together around 1670, close enough to Flamsteed's sighting to add credibility to it. Cas A still harbors a compact X-ray source near its center, probably a black hole or neutron star left from the core of the star.

We originally proposed to observe Cas A as a candidate for a pretty picture that could be released to the public to show how well Spitzer was working. It did not make the cut, so we converted it to an engineering observation to see how our instrument behaved when the detectors were scanned over a very bright object. The result is shown in Figure 26.6 (left panel). The supernova remnant is so bright that it drove some of our detectors off the scale (and in recovering useful information we satisfied our engineering objective). However, along the track of the scan were two blobs of filamentary material roughly equally distant on either side of the remnant. The rectangular scan geometry is fixed, but the direction is not. If our scan had been taken a few weeks later, we would have missed these blobs. They look as if they were shot out of the remnant, leaving trails traced out in additional filaments.

These blobs are at a huge distance from the supernova site; the angle from one to the other on the sky is about the diameter of the full moon. At the distance of Cas A, ten thousand light-years, the blobs appear to be fifty light-years removed from the explosion site. We had a problem: if the blobs had been launched at the time of the explosion, to get so far away they would have had to travel at about 15 percent of the speed of light! Therefore, we proposed that the material had been ejected by the massive, dying star long before it finally exploded. It could then have made its way out to where we detected it at a more conventional pace for clouds of gas and dust. Nonetheless, we were intrigued and took a picture with a large ground-based telescope at 2 microns. It showed the brightest of the blob filaments and emphasized their unique twisted fine structure. Four months later, German colleagues took another picture—and the blob structure had changed! In fact, the changes appeared to have occurred at nearly the speed of light!

The filaments were much brighter and easier to trace in the Spitzer 24-micron images than from the ground. Therefore, we hurriedly requested extra time with Spitzer to repeat our scan, and it was squeezed into the sched-

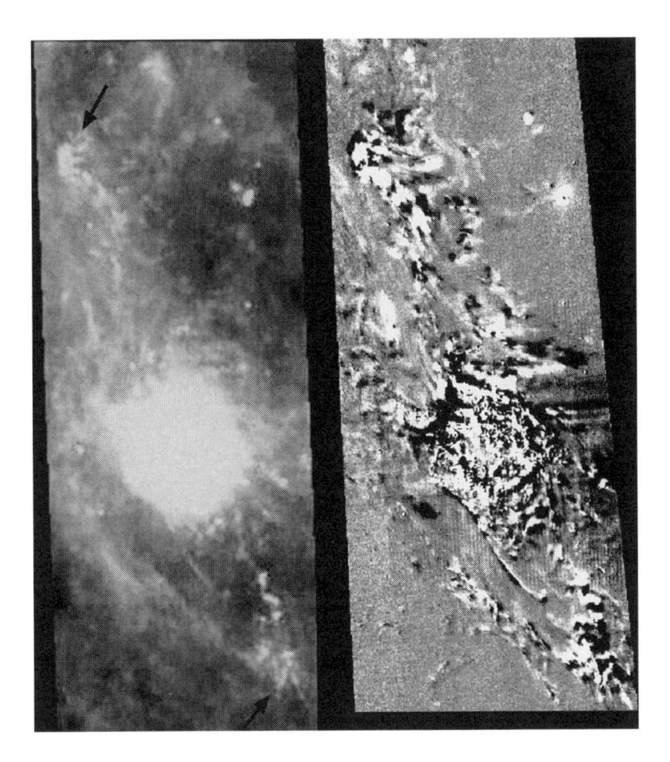

Figure 26.6. Supernova Remnant Cassiopeia A (Cas A). (Left) The "engineering" scan obtained in late November 2003. The supernova remnant itself is burned out below the center of the scan. Symmetrically above and to the left and below and to the right are two curious filamentary complexes of emission (indicated with black arrows). (Right) A difference image, subtracting a scan taken in December 2004 from the one shown at left. (O. Krause, personal communication, 2005)

ule just a year and a few days after our first observation. Huge movements had taken place, again at apparent speeds approaching the speed of light. The right panel of Figure 26.6 shows the difference between the two images. Areas that were brighter in the newer image appear black, while those brighter in the first image are white. Because in the difference image the complexes tend to be black away from the supernova remnant, the blob filaments were farther away in the later image. That is, they appear to be moving very rapidly outward from the remnant.

As with any true discovery, we had no premonition that such filaments could exist, and when we saw them we had no idea what was going on. We were wonderfully dumbfounded! After months of vigorous head-scratching, we have decided they are blobs of interstellar dust being heated by a blast of energy that is leaving the supernova at the speed of light; the blobs themselves are probably nearly stationary. When was the blast created? Perhaps it is from the original explosion in the seventeenth century. However, some aspects of the motions suggest it might be more recent, perhaps due to an explosion at the neutron star in the center of the remnant that might have occurred only about fifty years ago. As we trace the motions throughout the mission, we will eventually be able to choose between these possibilities.

APPENDIX A

Technical Aspects of Spitzer

A few simple principles of physics lie behind the Spitzer concept. Many of them revolve around temperature. In everyday life, we use the Fahrenheit and Celsius scales, which are based on readily obtained temperatures such as the freezing and boiling points of water. More fundamentally, temperature refers to the speed of motions of the atoms and molecules in a substance. These motions stop (within the limits of quantum mechanics) at the lowest temperature possible, called "absolute zero." The Kelvin scale, used by physicists, places zero at this lowest possible temperature and then adopts the size of a degree from the Celsius scale—that is, at 1 percent of the temperature range between freezing and boiling water. Zero Kelvin is −273 degrees Celsius or −454 degrees Fahrenheit.

Many laws of physics are much more simply expressed with the Kelvin temperature scale than with Fahrenheit or Celsius. For example, all objects emit light, or more properly electromagnetic radiation, which includes gamma rays, X-rays, ultraviolet, visible light, infrared, and radio. These varieties differ only in the wavelength of the electric field oscillations that power them (running from very short to very long in the order listed). According to the Wien Displacement Law, the maximum emission from an object is at a wavelength (measured in microns) of 2900 ÷ T (in K). Visible light has wavelengths ranging roughly from 0.4 to 0.7 microns. For example, for the sun, which has a temperature of about 5600K, the maximum output is right in the visible, 2900 ÷ 5600 = 0.518 microns. Room temperature is near 290K; a ground-based telescope is also close to this temperature and has maximum output at 2900 microns ÷ 290 = 10 microns. This wavelength is twenty times that of visible light, placing it well into the infrared. Thus, a ground-based telescope emits hardly at all in the visible range but brightly in the infrared. In the latter spectral range, astronomical sources must be detected against the bright glow of the telescope, a challenge that has been compared with finding a match in a blast furnace.

The power emitted by an object is proportional to its temperature in Kelvin raised to the fourth power, T^4 (i.e., T times T times T times T). Thus, cooling the telescope to make it darker in the infrared has two benefits: it moves the peak of the emission to longer wavelengths (according to the Wien Law) and it also rapidly reduces the total output. Unfortunately, any significant cooling of a ground-based telescope has the disadvantage that gases from the atmosphere will condense on it, first water and then others if one stubbornly keeps forcing the temperature down. A frosted-up telescope is useless because the frost keeps light from reaching the detector. The only solution is to put the telescope in a vacuum where there are no gases to condense—that is, to put it into space. In the case of Spitzer, cooling to 6K moves the peak of the emission to about 480 microns, long enough that even the longest-wavelength band in Spitzer, at 160 microns, cannot see it. The detectors of a telescope at this temperature are limited only by the faint glow of the warm dust in the zodiacal cloud around the sun, which, like a ground-based telescope, is near 290K in temperature. Unlike a telescope, the zodiacal cloud is very dilute, with dust grains covering less than one-millionth of the sky. Hence, its emission is a million times fainter than that of a telescope at the same temperature. The noise of a high-performance detector is proportional to the square root of the strength of the signal falling on it. Thus, all other factors being equal, a cold telescope in space can detect objects one thousand times fainter than the same telescope could detect on the ground.

The low temperatures required for Spitzer are maintained with liquid helium, which boils at 4.2K when it is at a pressure equivalent to that of the atmosphere of the earth at sea level. When the helium absorbs heat, it maintains its temperature by boiling, just as water absorbing heat from a stove boils to maintain its temperature. In a "cold launch" observatory, such as ISO (Figure A1), the liquid helium is put into a vessel surrounding the telescope and cools it directly. To prevent boiling the helium away too quickly, both vessel and telescope are suspended with supports that

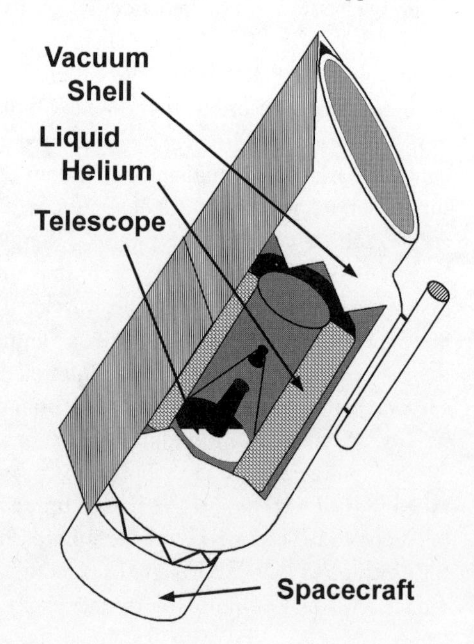

Figure A1. Infrared Space Observatory (ISO). The observatory is 5.3 meters long and weighs 2400 kilograms. The primary mirror is 60 centimeters in diameter. (After a figure from the European Space Agency Web site, http://www.iso.vilspa.esa.es, 2004)

conduct as little heat as possible in the middle of a carefully designed vacuum-tight container, out of which all the air has been pumped. This device is the dewar (named after its inventor, Sir James Dewar, but similar in operation to a common thermos jug); it minimizes the heat entering the liquid helium because there is no air to conduct the heat from its outer walls to the helium vessel. The instruments are mounted on the back of the telescope, with thin wires to bring electricity to them and carry their signals out. Before launch, the helium vessel is filled, so the telescope and instruments are cold. Once on orbit, a vacuum-sealed and helium-cooled trapdoor on the front of the dewar is opened to allow light into the telescope.

Frank Low realized at our Broomfield retreat that we could use a different approach by taking advantage of a general property of most materials: when they are already cold, the amount of heat that must be removed to reduce their temperature further is greatly reduced. To take advantage of this behavior, Low proposed that the instruments be mounted inside a helium dewar as before and launched cold. The telescope, however, would be mounted outside rather than inside the vacuum-tight shell of the dewar (Figure A2).

We can describe the other parts of the full observatory depicted in Figure A2 by

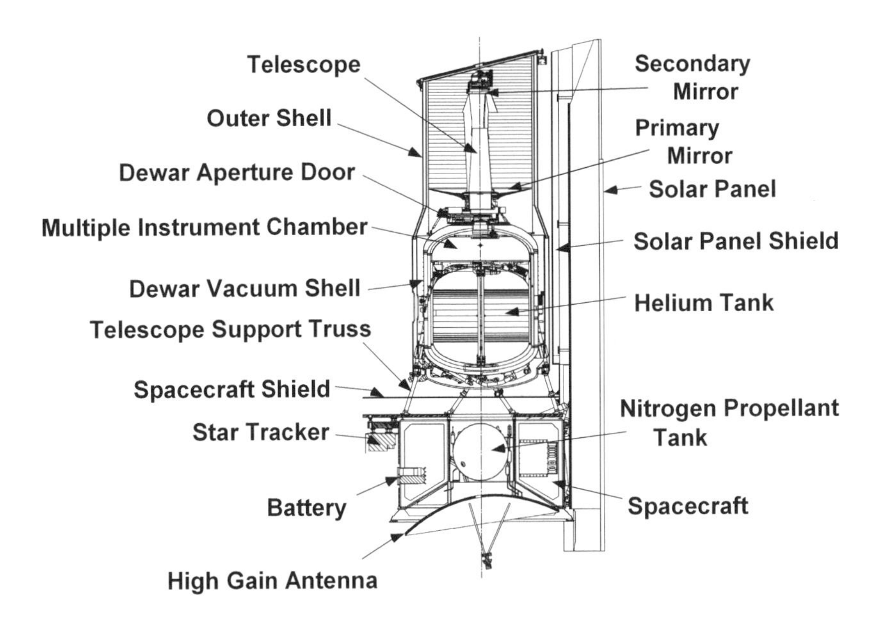

Figure A2. Spitzer Observatory. The figure illustrates the central features of the warm launch architecture (by comparison with Figure A1), as well as the usual support features for any space observatory. The observatory is about 4 meters long and weighs about 865 kilograms. The primary mirror is 85 centimeters in diameter. (Published on the Spitzer Science Center Web site, http://www.spitzer.caltech.edu, 2004)

tracing the fate of an infrared photon as it is converted into an electronic signal and sent to Earth. Our photon enters the telescope, traveling downward from the top of the figure. It is reflected off the primary mirror to the secondary mirror, and from there past the dewar aperture door (presumed to be open) into the multiple instrument chamber (MIC). The telescope is a traditional reflecting design that collects light with a concave primary mirror and relays it back to our instruments with a convex secondary mirror (a "Cassegrain" system, named after its seventeenth-century inventor).

The most fundamental limitation of a small telescope is angular resolution, the level of detail in images. The achievable resolution becomes worse in direct proportion to the wavelength of the observations, and inversely with the mirror diameter. Since Spitzer was to work at relatively long wavelengths, many of our potential comparisons with observations by other facilities were going to be limited by our poor angular resolution. Thus our efforts to keep the telescope from shrinking during the entire history of the project, retaining a modest 85 centimeters (34 inches) of primary mirror aperture.

The MIC is attached to the top of the liquid helium tank inside the dewar vacuum, where the helium cools it directly. We assume that the telescope has been pointed to bring our photon to the entrance of one of the three instruments within this chamber. During the twelve years of design and construction we had been forced to shrink the instruments a lot: each started out the size of a wardrobe and ended up more like a hatbox. Along the way they lost many capabilities such as mechanisms to bring many filters into the beams of the imagers, and even whole instrument sections such as a far infrared spectrometer and a photometer operating in the submillimeter range. They are now very lean, purpose-built around the needs of the four science themes. Thus, the Infrared Array Camera (IRAC), under the leadership of Giovanni Fazio, has just four fixed color bands at which it observes, each band being imaged onto its own array of detectors. The bands just cover the 3.6–8-micron range needed to study galaxies out to high redshifts. It happens that one of these bands is also ideal for detecting brown dwarfs, since they have a transparent region in their atmosphere near 4.5 microns where much of their internal heat escapes. The Multiband Imaging Photometer for Spitzer (MIPS), my instrument, observes in just three color bands from 24 to 160 microns. Since three colors seemed too few to cover a factor of seven in wavelength, we added another mode where the light from a small field on the sky is spread in color to provide about a dozen bands between 50 and 100 microns. The three fixed bands in MIPS are optimized around the needs for deep surveys to measure the far infrared properties of distant galaxies and for studies of debris disks, while the greater color discrimination was provided primarily for debris disk observations. The Infrared Spectrograph (IRS), led by Jim Houck, takes the form of four separate mini-instruments that together cover the 5–38-micron region at low spectral resolution (dividing the light into about a hundred color bands over this entire range) and also the 10–37-micron region at moderate resolution (dividing the light more finely, into about a thousand narrower bands). When displayed in a continuous range, like the colors of a rainbow, these color patterns are called spectra. The first capability is used to determine the broad distribution of the output of a source over colors. Since different types of complex molecules and minerals have characteristic broad patterns of bright and dark in their spectra, the low-resolution part of the IRS provides a critical capability for understanding the structure and composition of debris disks, particularly if supplemented by the longer-wavelength capabilities of MIPS. It is also our means to refine estimates of the redshifts of galaxies that do not have features detectable with large ground-based telescopes, since interstellar dust emits in a characteristic set of broad spectral features. The second capability is designed around detection of emission lines, typically the output of single atoms or relatively simple molecules narrowly concentrated at precise wavelengths associated with their internal energy states. It is our means to find active galactic nuclei that are so buried in dust that the characteristic optical emission lines cannot escape, and to study other emission lines such as those from dust-embedded star-forming regions in both our own galaxy and ultraluminous galaxies.

An explanation of how the other two instruments were built would agree in outline with the account in Chapter 13 of how my instrument was built. Rather than provide parallel accounts, I will instead describe the unique design challenges faced by the other teams.

The spectrograph should have been the biggest challenge under the Spitzer minimalist instrument guidelines. Motor-driven optical elements are usually employed to adapt a spectrograph detector to the full range of wavelengths and levels of color discrimination ("spectral resolution") provided by the instrument. However, Houck realized that it is the *first* detector array that might be expensive enough to justify this level of complexity. Once the manufacturer has learned to make one, additional arrays can usually be provided for a relatively small marginal cost.

Therefore, he had the IRS built as four separate spectrographs using four detector arrays. Even though this approach would seem to increase the complexity, it has the advantage that each spectrograph can cover a given wavelength range with no moving parts. Two of the spectrographs provide low-resolution spectra (with color discrimination [spectral resolution] of about 1 percent of the observing wavelength) from 5 to 38 microns. The two others provide spectral resolution about ten times greater.

Each of these modules operates on similar principles. First, the light must be sorted so that different wavelengths will not be confused after they are spread into a spectrum. To do so, the light passes through a slit and filters. The light that remains proceeds to a mirror that converts it to a parallel beam (the collimator mirror), then to a grating that spreads the different colors into different directions, then to a "Schmidt plate" that adds subtle optical corrections, and finally to a camera mirror that focuses it onto the detector array. A peakup channel is similar except it uses a flat mirror in place of the grating, so that it forms images of a small region of sky. A source is first placed in the peakup field and its position determined, and then the telescope is offset precisely to put it in one of the four IRS spectrographs' slits.

The challenge with the Infrared Array Camera was to make it small. Under our minimalist approach, IRAC was not to use a moving wheel to put different spectral filters into the optical beam to allow observation at different colors. Instead, it too was built around four detector arrays, each to provide a specific fixed color over a large imaging field; the bands are at 3.6, 4.5, 5.8, and 8 microns. Although construction of the instrument would have been easier if it had used mirrors for all of its optics, such a design would have been too large; IRAC uses lenses to focus the light onto its detectors. Since the four fields would not have fit into the telescope's focal plane, IRAC divides the light from only two fields in the focal plane into the four colors. After the light from one of these fields passes through a lens, it encounters a "dichroic," a spectral filter designed to reflect one range of wavelengths and transmit an adjacent range. Thus, a single field in the telescope focal plane is brought to two detector arrays observing two spectral bands.

Taking the Hubble Telescope as a model, a typical astronomical instrument would have three or four motor-driven mechanisms to change the optical components in the beam, resulting in a couple of dozen spectral bands for an imager and a variety of spectral resolutions and wavelength ranges for a spectrograph. Our imagers have two mechanisms between them and only seven spectral bands. Our spectrograph has no moving parts at all; it is basically the equivalent of a box camera that we just aim and shoot. The ingredient that makes these limited capabilities so powerful is the arrays of detectors.

After a photon is "processed" optically within an instrument, it impinges on one of these detectors, where it is absorbed to free an electron that can migrate through the detector volume. The electron is collected onto a very sensitive transistor and the signal is amplified. The resulting still very small electric current is conveyed by delicate wires out of the dewar and down into the spacecraft. There, a more power-hungry set of electronics further amplifies the current and converts it to a digital, computer-compatible signal. After additional processing to organize the data, it is stored in the large digital memory in the spacecraft computer. When the time comes to send the data to Earth, the observatory is maneuvered to point the high-gain antenna in that direction, and a Deep Space Network (DSN) antenna is directed toward Spitzer. The spacecraft then broadcasts the contents of its digital data memory through the high-gain antenna to the DSN.

The outer shell protects the telescope from photons arriving from unwanted directions, helping isolate the signal from the object of interest. The dewar aperture door was closed for operation on the ground and during launch to keep air from entering the dewar vacuum. It was opened only after the observatory had been delivered to space. The solar panels generate the power for the spacecraft, and the battery provides continuity in case of short interruptions. The star tracker is the heart of the system that points the telescope. The pointing is stabilized by a combination of reaction wheels (not shown in Figure A2) and a compressed gas reaction control system of small jets powered by the tank of nitrogen gas in the center of the spacecraft. The telescope's outer shell was carefully blackened to allow it to radiate thermal energy into space and cool the telescope. The remainder of the cooling is supplied by the cold gas escaping from the dewar. Thermal shields protect the cold telescope from heat radiated by the warm solar panels and spacecraft.

When it got on orbit, the telescope initially radiated efficiently to cold space (Figure A3). This "radiative cooling" brought its temperature down rapidly. The efficiency of this process drops rapidly with falling temperature, however, in proportion to T^4. By the time the telescope reached 40K, the amount of radiation it was emitting, and hence the radiative cooling power, had been reduced by a factor of about three thousand. By then the helium could absorb the remaining heat, which was removed by piping cold helium gas from the boiling liquid through heat exchangers connected to the telescope.

During the twenty years between its development and launching, Spitzer was kept viable scientifically by improvements in arrays of infrared detectors. The approach to manufacturing such arrays had just been invented at Hughes Aircraft when we wrote our instrument proposals in 1983. It involved manufacturing a grid of low-noise amplifiers on a wafer of silicon and a matching grid of infrared detectors on a

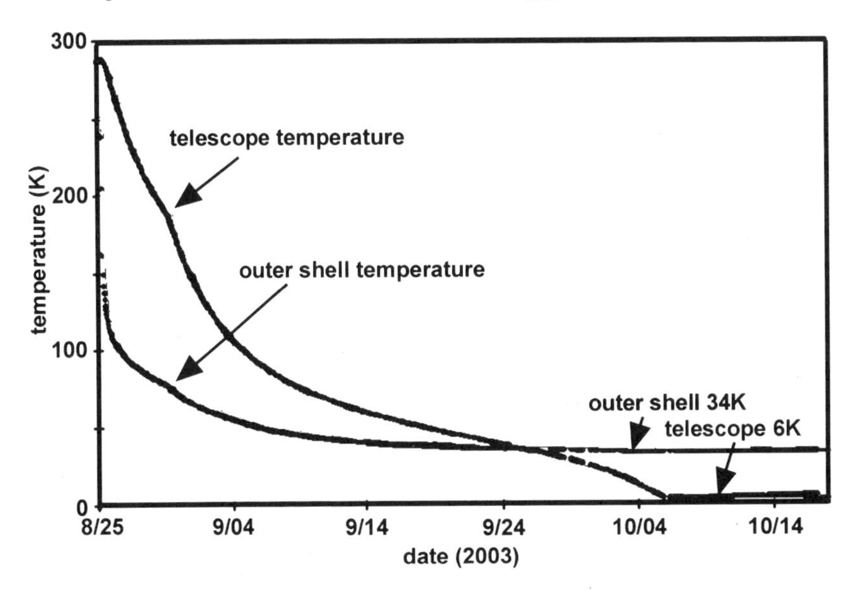

Figure A3. On-Orbit Cooling of the Spitzer Telescope. (P. T. Finley, personal communication, 2003)

wafer of material suitable for them. The detectors convert photons to electrons and the amplifiers boost these tiny signals enough to send them to conventional electronics. The two grids were manufactured with matching contacts, and a tiny bump of indium was deposited on each one. The grids were then carefully lined up and squeezed together. When the indium bumps came into contact and were distorted by the pressure, the indium cold-welded and established the electrical contacts between detectors and amplifiers (Figure A4). This approach allowed the detectors and amplifiers to be built on different materials and to follow different processing steps, each optimized for its own application.

Fortunately for astronomers, the immense cost of developing detector arrays operating from 1 to 28 microns was paid by the military. We could simply work with the manufacturers to push their performance to a higher level. Increasingly high performance infrared detectors have been made possible as capabilities have improved for growth of multiple-layer semiconductor devices with extremely good control of impurities. Commercial interests in complex integrated circuits paid for the facilities required to improve the array readout chips. The result is that the arrays used in Spitzer far outperform their military-based ancestors. For the longest wavelengths, though, we were on our own, and we had to develop arrays using a very different set of construction techniques.

The maturation of infrared array technology for the 1–28-micron range and the

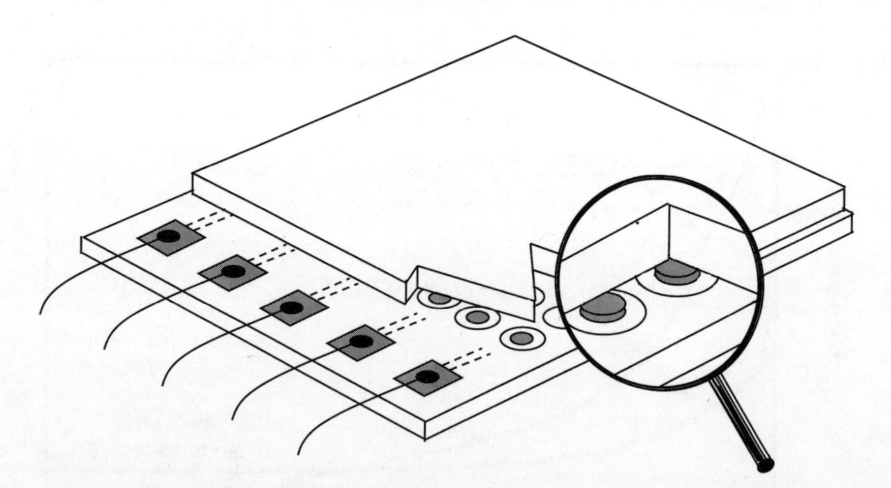

Figure A4. A Hybridized Infrared Detector Array. The detectors are on top, and a silicon readout integrated circuit wafer is on the bottom. The readout is provided with voltages and clock signals through wires attached to bonding pads, and its output signals are delivered to external circuits in the same manner. The detectors are attached to the readout amplifiers by evaporating an indium bump on each detector and on its mating amplifier, and then aligning them and welding them by pressing them together.

development of a new technology for the longer wavelengths occurred in parallel with our long, frustrating wait to get Spitzer under way. Thus, as Spitzer languished waiting for funds to get started, infrared arrays quickly grew in number of detectors and the amplifiers were reduced in noise. The largest arrays in the Spitzer instruments have sixty-five thousand detectors, and their amplifiers are quiet enough that signals of about ten electrons can be detected. This performance is a huge advance over that available when the Spitzer teams were first selected, when the most advanced arrays had about four thousand detectors and could only detect signals of a few hundred electrons (just a few years prior to that, no infrared detector arrays were available at all). According to an overall performance metric for estimating the relative power of detector arrays of differing size and amplifier noise, the arrays launched in Spitzer are about ten thousand times more powerful than the ones available in 1983. These advances in detectors accounted in large part for the ability to provide a powerful, general-purpose mission on the budget that was available for Spitzer.

APPENDIX B

What Spitzer Demonstrated about Building Space Missions

Struggling against funding limitations is almost as common in space science as dealing with technical problems. For example, the Hubble Space Telescope team had to agree to drastic budget reductions to get congressional approval, reductions that were to be made possible through streamlining development procedures. It quickly became apparent that the assumptions were not realistic, resulting in the explosive budget growth in the early 1980s (Smith et al. 1993). Chandra was descoped during the budgetary panic in the early 1990s (Tucker and Tucker 2001). Smaller projects that compete within NASA must promise almost unbelievably good performance for the cost to be selected, with the same final result. Setting aside the initial twelve years of downscaling pressures on SIRTF, once we actually got started, the funding pressure on us came in the form of the faster, better, cheaper guidelines then in force. In this appendix I discuss how the project management team coped with this challenge.

"Faster, better, cheaper" is a slogan, not really a management philosophy. As part of the reaction to the failures in 1999, Tony Spear, who managed the Mars Pathfinder—the "faster, better, cheaper" icon—led a task force to define the slogan (Spear 2000). It was to be viewed as an effort to extend the innovation and efficiencies that were being realized in the lowest-cost NASA missions upward into larger, more complex, and costlier ones. The task force concluded that the system required, among other things:

- latitude to adjust mission scope to fit within the cost cap
- stable requirements and reliable funding
- work in teams cutting across institutional boundaries
- increased reliance on internal procedures and personal responsibility to intercept errors and build in quality
- thorough engineering and strict project management (i.e., no relaxation from standards for the mission foundations)
- follow-through on details
- constant assessment and mitigation of risk
- extensive peer review and informal interactions with experts outside the project

The first wave of faster, better, cheaper missions was felt by Spear's group to have been given reasonable goals within their cost caps, and hence had had a reasonable chance of success. In fact, they were all largely successful. For the later missions, there was an attempt to get even more for fixed costs and within rapid schedules, creating goals that were unrealistic. Many of the failures had roots in this difficult challenge; for example, the omission of a system-level test on Deep Space 2 (a part of Mars Polar Lander that failed independently of the main mission) and the design of Mars Polar Lander with no telemetry during the key descent stage. Other failures, such as WIRE and the Mars Climate Orbiter, seemed to arise from a relaxed follow-through on details that was contrary to good practice but was perhaps inadvertently encouraged by the faster, better, cheaper culture.

Brian Muirhead, the Pathfinder flight system manager, explained the difference as follows: "One of the keys to managing risk . . . was understanding where the risks were very early in the project, and then developing plans to mitigate them. . . . We took risks by accepting new ideas but then made sure they would be successful. When you have a standard way of doing business and somebody says, 'Take more risk,' there's a good chance you'll try to do that by cutting corners on the same practices that have worked in the past. You'll probably fail" (Muirhead and Simon 1999).

It was recognized that the emphasis on teaming and individual responsibility would become increasingly difficult for projects above a certain scale, placed roughly at $200 million. The issue is not just one of scale, but also of change in organization. Within NASA, small projects can be conducted largely at a single center. Large projects are nearly always carried out through an alliance of the responsible NASA center with a number of contractors, each with its own management team. A systems management umbrella traditionally coordinates these efforts. The faster, better, cheaper emphasis on team relationships rather than systems management weakened this coordinating umbrella.

Spitzer exceeded the $200 million threshold by more than a factor of two and included major contracts to Lockheed, Ball, and the three instrument teams. Nonetheless, our initial management approach incorporated all of the items listed above save the first. In common with late faster, better, cheaper missions, we had also been strongly encouraged by NASA Headquarters to squeeze more content into the mission than our costing exercise indicated was prudent.

Simmons's initial approach to building Spitzer, in common with the ideas behind faster, better, cheaper, emphasized developing a new projectwide culture—a set of shared values among the project participants that are felt at a level where they are generally accepted without being stated. Simmons established a multiorganizational team whose members were selected on the basis of their capabilities rather than on their response to a predetermined set of guidelines and questions (as they would have been with a traditional selection based on a detailed Phase C–D proposal). It was assumed that the members of the team shared common goals and a common vision of how to reach them, so they could be empowered to carry out their responsibilities without continuous systems management supervision by the project. When he came onboard, Gallagher continued the teaming approach. Simmons had been his mentor on a previous instrument development for HST, so he felt at home: "I was very comfortable with the culture," he said at one point. "I have experience with another culture that is very strict, driven by contracts and change orders. It works, but it is very slow."

One key issue in applying this approach to projects on the scale of Spitzer is that each participant comes with a preexisting organizational culture that must be melded into the general one.

The university-based science team–led developments were very compatible with the faster, better, cheaper Spitzer culture. The flexibility and innovation bred by years of developing ground-based instruments on small budgets complemented the discipline and thoroughness of the aerospace engineers to produce an efficient yet rigorous environment for spaceflight instrument design and construction. Equally important, the scientists at the universities had proposed their involvement in 1983, and then had waited twelve years for even a taste of a reward. Spitzer was likely to be nearly their last hurrah professionally, and they made success their highest priority.

In this context it is probably significant that the third instrument, IRAC, experienced difficulties in holding to budget and schedule. Because Goddard would not permit the principal investigator–led team to have in-line authority but would only accept official direction from JPL as another NASA center, the Goddard program was insulated from the urgency and focus felt by the university-led instrument teams. With its convoluted and diffuse line of authority to the instrument team, IRAC also was an easy target for programs within Goddard needing to find personnel or center resources for their own needs. Keeping key personnel was a continuous challenge.

These issues were compounded by the conservative project culture characteristic of Goddard programs. Many of the skilled personnel at this center were paid as civil servants and hence were available without substantial cost to a project (the situation is changing by implementing "full-cost accounting" at the center). As a result, efforts at Goddard were partially shielded against demands for greater efficiency such as those embodied in faster, better, cheaper. In contrast, JPL had no civil service positions and thus had to respond fully to such initiatives. On top of this basic difference, Goddard and JPL are the two main science instrumentation centers within NASA and as such are fierce rivals. JPL was therefore uniquely poorly positioned to impose new management concepts on Goddard.

The teaming culture also encouraged the Spitzer scientists to maintain the strong involvement in the overall mission that had been a hallmark of the project from their selection. Simmons welcomed this relationship: "My background was in instrument development, which made me think of Spitzer as a large instrument with necessary support from a spacecraft, not as a spacecraft that happened to carry some instruments." Gallagher later described the results: "The way the scientists were involved was a totally unique aspect of the project and very positive. When we had a problem, I learned to look to the scientists at least as much as the engineers or managers to find a solution. On other projects, the scientists tend to show up at the beginning to set requirements and then wait impatiently for their data."

The aerospace members of the Spitzer team present an interesting contrast in culture. NASA space programs are generally run in a very open fashion, with a free flow of ideas that circumvents distinctions of rank and with public reviews at every stage. The result is that many of the engineers and managers on a NASA project have a broad perspective on its goals. In contrast, military space development is usually conducted under the constraints of secret classification and the resulting mandatory impediments to free communication. Also, the military customer is rigidly organized in a strict hierarchical structure that imposes its own restrictions on interactions with a contractor.

These different attitudes can dominate the corporate culture. Ball Aerospace has consolidated its NASA and other civil space efforts onto a campus in Boulder, Colorado, where the open style of operation is dominant. Virtually every engineer is involved in a project across a variety of its aspects and is knowledgeable about its ultimate goals as well as about the immediate challenges. In addition, Ball is the dominant U.S. aerospace supplier in many areas of space science. There is a continued flow of activity through the Boulder campus, providing potential paths for career growth within this style of business.

When it proposed for its role in Spitzer, Lockheed Sunnyvale had a similar group focused on civilian projects, in Building 107 at one corner of its large plant. Soon after that, however, with the company under pressure to improve efficiency and downsize its workforce, this group was integrated into the rest of the Sunnyvale effort. Where the NASA business is a small portion of the total in a plant dominated by secret defense work, as it was at Lockheed Sunnyvale after this reorganization, a more isolated mentality naturally rises to the top. Poor communication among engineers working in different areas on a "secret" project can become a requirement rather than a problem. Workers on such projects are not supposed to be told about other areas unless they have a "need to know." Thus, communication across engineering disciplines is discouraged. Moreover, since most of the corporate bills are paid by the defense contracts, those working on them often come to consider themselves an elite upon whose efforts the health of the company depends. The military customer is comfortable interacting only with this elite, since it is also a strictly hierarchical organization. The corporate elite are marked by security clearance, which qualifies them uniquely for the "dark-side" projects. It is often felt that obtaining a clearance is in itself worth substantial additional income over the long haul, since it qualifies one for work assignments and advancement paths that are otherwise inaccessible. Once obtained, a clearance seems to be wasted by working on unclassified NASA projects. The mandatory secrecy and elite-based structure can also become excuses to treat NASA projects with disdain when someone begins to feel that his/her inside knowledge provides insights that cannot be communicated to others. This culture is likely to be resistant to the teaming philosophy required for coordination in a faster, better, cheaper management structure. Fortunately, there were some personnel at Sunnyvale who were attracted to Spitzer because they did not like the military-based culture. They formed the core of our effort there.

Project morale is important in any situation. The empowerment that was part of faster, better, cheaper boosted morale in the instrument teams and at Ball because both groups were culturally compatible with the approach. However, morale within Lockheed was often low, as can be judged by the handling of the fee. Though the fee may be a small part of the budget for an aerospace firm, it is the metric against which management is judged by stockholders and investors, and thus it assumes a disproportionate importance. To take maximal advantage of this focus, aerospace contracts often use an "award fee" in which the final value is based on the contractor's performance in carrying out the plan and overcoming unexpected problems, as graded on an elaborate report card. In exasperation, JPL eventually tried to express its displeasure with Lockheed's overall performance by reducing the award fee. Normally, a contractor would have accepted the challenge to improve performance and win a larger fee next time around. Instead, a wave of negative reactions swept through the Lockheed team, driving morale still lower and further reducing performance. Eventually, JPL gave up the award system altogether because it seemed to be counterproductive.

Having discussed the members of the Spitzer team, we now turn to its overall performance in terms of the usual management metrics of cost and schedule. The usual way to evaluate the cost performance of a mission is to compare the as-built cost with the initial estimate. By this measure, we planned to build Spitzer for $458 million, and we actually spent $666 million, a growth of 45 percent. There are other ways to look at the situation, however. If we had taken everyone's "not-to-exceed" numbers presented in May 1997 just before the "June deal" and added a 30 percent contingency, the result would have been $510 million for the estimates and $153 million for the contingency, virtually identical with what was actually spent.

In common with most second-generation faster, better, cheaper missions, we were pushed toward a project that was overextended compared with a conservatively estimated budget, making an overrun inevitable. When that occurred, the rigid faster, better, cheaper approach was no longer in vogue and NASA was willing to find more money. In parallel, NASA established additional review panels and took other steps that drove us into a more traditional, conservative management style. Only IRS and MIPS were truly developed and delivered under faster, better, cheaper guidelines; the remaining project elements worked according to increasingly traditional rules.

We now consider schedule. The launch date slipped by about seventeen months during the development of the mission (not counting the four involuntary months tacked on at the end). Nominally, most of this slip is attributed to the late delivery of IRAC, which held up integration of the telescope, and the delays in the spacecraft, particularly of its software, which held up integration of the observatory (telescope and spacecraft). In fact, however, the telescope team would not have been ready if all the instruments had been delivered exactly on time. Given the pace of the telescope, IRAC may be responsible for one to two months of delay, no more. The telescope was delayed further by the dewar overpressure and the need for a second test in the Brutus chamber. As a result, the telescope could not have been delivered more than one or two months earlier than it was needed for integration with the spacecraft, on its delayed schedule. The spacecraft hardware had multiple problems that would have surfaced as schedule delays if the software had been ready even slightly sooner. There were further delays in the observatory integration and testing, but the mission operations effort was also struggling to make the deadlines required to operate the observatory in space and without which we would not be able to launch.

The conventional view of delays is that the group that is furthest behind gets blamed for all of the delay that results, allowing those who are in only slightly better shape to escape nearly blameless. Thus, IRAC and Lockheed took the blame, but in fact we needed an extra year added to our initial aggressive faster, better, cheaper schedule for many reasons.

Virtually every NASA Space Science mission ends up exceeding its initial budget and schedule, so at some level this behavior is built into the system. The competitive forces encourage projects to minimize their budgets, often by making optimistic assumptions about aspects of the mission that come late in the development and hence are not planned in detail. The cost models never catch up with this bias because they do not account for the politics of getting a mission started. Our vast underestimation of the needs for mission operations is an example. Eventually NASA Headquarters has to find the money to complete the mission (or take the painful step of canceling it and losing most of the investment already made). A better measure of the cost performance of a mission is to compare its growth with historical trends (Roy 1998). The distribution shows that 30–60 percent growth is typical of well-managed missions. Missions that encounter real problems lie in an extension of this distribution upward to factors of two and three.

Despite the eventual increase in budget, the cost cap and the resulting new management approaches brought benefits to NASA. Inflating the mission costs to a common basis, HST was more than five times more costly than Spitzer, Chandra cost about three times more, and Compton Gamma Ray Observatory cost about one and a half times more. These ratios apply only up to launch and commissioning; the long string of servicing missions to HST adds substantial additional cost. None of the ratios includes the launch costs. Since the other three missions relied on the very expensive space shuttle while Spitzer used a relatively cheap Delta rocket, the total differences are larger than listed; even the Gamma Ray Observatory exceeds twice the total cost for Spitzer.

Since projects behave somewhat like households—the larger the income, the more gets spent—with a higher cost cap it seems likely that Spitzer would have cost more. It is possible to estimate this effect since the Atlas version of the mission unveiled in 1993 represented a serious effort to cut the mission cost, but without the full late-phase faster, better, cheaper–based culture. We can assume (with more than a little optimism) that it could have been built for the estimated cost of $1 billion, inflated to $1.1 billion to place it in the appropriate time frame. Much of the losses in the original science capability of the mission had taken place to get to the Atlas version. In addition, the change in concept from cold launch to warm launch probably did nothing substantial to bring the mission cost down. Although the lower mass of the new concept enabled use of a smaller rocket, the rocket costs are not included in the numbers above and this effect can be ignored. Thus, the perspective of the Atlas mission suggests that the Spitzer cost would been more than 50 percent higher if it had been built under the old rules and with a higher initial budget.

How far can this process of savings by applying cost pressure go? Is "faster, better, cheaper: pick any two" just cynical humor, or does it express a deeper truth? The range of management styles applied to Spitzer makes it an interesting case to probe this question.

David Bearden of the Aerospace Corporation examined issues such as these with systematic studies of the performance of satellite missions (Bearden 2001). He converted a number of spacecraft parameters like mass, distance from Earth, pointing accuracy, data rate, and number of instruments into a single numerical complexity index. He then plotted various other development parameters against this overall indicator of complexity. With regard to "faster," he found a dramatic clustering of failed missions in the zone of high complexity and short development time. This tendency seems to override all other factors such as adequacy of budget. That is, if the schedule is too short, even budgetary largess cannot overcome the risks. Not surprisingly, there is a similar clustering of failed missions in the zone for high complexity and low cost, the "cheaper" zone.

What about "better"? What, exactly, does it mean in this context? In the original demands for a faster and cheaper management style to make a manned Mars program plausible, faster and cheaper by themselves had been considered to be "better." To be less ambiguous in the later use of the same trilogy to describe a downsizing of science missions, this word was sometimes replaced by "smaller" (Roy 1998). However, small spacecraft are not necessarily less expensive than large ones to accomplish equivalent goals (Sarsfield 1998). Multiple parallel developments tend to have high overheads compared with a single large project. A better substitution would be "simpler," indicating a desire to use advanced technology to drive down complexity. Traditionally,

NASA has tended to be very cautious about using new technologies, under the belief that mission failures would result. However, after-the-fact studies indicate that this risk is often overrated, and that new technology has not been a major contributor to mission failures (Sarsfield 1998).

With time, however, "better" came to indicate that a high-launch-rate, slightly-higher-risk program would provide better science opportunities. The premise was that the risks would have only minor effects on overall mission success. To test this premise, we can look at the areas that have proven failure-prone in the past.

Of seventy-one missions listed on the NASA Space Science Web site prior to the Spitzer launch, the dominant cause of failure (seven of thirteen) was either in the launch or in later rocket firings, and in spacecraft control. The launch problems were largely independent of the management rules for the payload. If there is a pattern to the remaining failures, it is that faster, better, cheaper and similar missions tend to have an elevated rate of problems with areas of conventional engineering that are specific to the mission but not at the core of its scientific instrumentation. Examples include the inadvertent cover release in WIRE, the physical units confusion in the Mars Climate Orbiter, and the software flaw in the Mars Polar Lander.

A huge amount of conventional engineering is required to make a satellite work. When such engineering has been tested by use in many satellites, it is generally very reliable. In areas where it is new and mission specific, because it is "conventional" it may not attract the same level of attention as the unconventional engineering in the high-technology items. As a result, these mission-specific areas are a recurring weakness in the defenses against failure.

Within Spitzer, the instruments were most in tune with faster, better, cheaper. Should we therefore have considered them to be high risk? Using the listing of past and current missions provided by the NASA Web site, virtually no mission has been totally lost due to failure of instrumentation. Frequently a mission carries a number of individual instruments, and the failure rate even of single instruments is very low. It is difficult to quote an exact percentage, since the definition of failure is a bit subjective—frequently instruments do not work exactly as planned when they get to orbit—but in this statistic I count a failure only if a large portion of the originally anticipated data is not returned. The actual performance of the Spitzer instruments is consistent with this historical trend.

A general attitude within NASA is that instruments are high risk because they contain one-of-a-kind new devices and push the state of the art in technology. In the case of Spitzer it is noteworthy that *all* of the new high-technology areas—instruments, telescope, and dewar—were completed without major difficulties (but not without many close calls!) except for obvious low-technology mistakes like the dewar overpressure incident and the failure of the first Brutus test. Our problems were in more conventional items: the spacecraft software, the IRAC firmware, the IRS and MIPS circuit cards, mission operations, the reaction control system.

Instrument construction falls within the $200 million rough upper limit for the faster, better, cheaper development approach proposed by Tony Spear's task force. Because the goals for an instrument are clearly defined and the team building it is relatively small, it is feasible to work in the culture of high reliance on individual responsibility, good communications, and effective use of internal procedures. And because NASA *suspects* that the instruments are the most failure-prone elements of missions due to the new technology they frequently contain, they are subjected to a rigorous review process. That is, although the faster, better, cheaper approach has been officially abandoned, it captures many of the principles of good management on modest programs or modest elements of larger programs. The experience with SIRTF also demonstrates the risk on large projects. For success, any set of management principles must be compatible with the cultures of all the project elements, and cultures within large corporations and other organizations cannot be changed easily.

Bearden's approach to assessing mission cost, budget, and risk submerges these details into an overall view. He ran a model of Spitzer costs shortly after the observatory had been launched (Bearden, personal communication, 2003). His conclusion was that our budget was well into the "safe" zone appropriate to our degree of complexity, and that our schedule was tight but adequate. Although Spitzer started under strict faster, better, cheaper guidelines, its cost and schedule growth eventually put it into more familiar territory for management of major NASA missions. The pressure to reduce cost did not create a breakthrough in overall efficiency, but it did allow Great Observatory complexity—high pointing accuracy, large data volume, deep space operation, and long life—to be squeezed into a compact, lightweight envelope.

NOTES

Chapter 1. Friday the Thirteenth

1. A micron is one-millionth of a meter; there are 25 microns in one-thousandth of an inch. The "electromagnetic spectrum" describes waves of electricity and magnetism with identical properties except for wavelength (the distance from wave crest to wave crest). It ranges from the radio (wavelengths of meters) through the X-ray (wavelengths of billionths of meters) to gamma ray. Visible light has a wavelength of about 0.5 microns, or 0.0000005 meter, so SIRTF is designed to detect photons with wavelengths from six to four hundred times that of visible light.

2. Kelvin temperatures, abbreviated K, are measured in a degree the size of that in the Celsius system, but with the zero of the scale set at "absolute zero." Absolute zero is the coldest achievable temperature, at which the thermally driven motions of atoms stop to within the limits of quantum mechanics. Every object warmer than absolute zero emits electromagnetic radiation, with a characteristic wavelength proportional to a constant divided by the absolute temperature. Thus, something at room temperature, 300K, emits mostly around 10 microns, while something near 6000K, such as the sun, emits mostly around 0.5 microns.

3. A dewar is a type of high-performance thermal flask for cold liquids. The liquid is placed inside a vessel that is suspended within a vacuum maintained within the dewar's airtight outer case. The insulating properties of the vacuum, along with careful design of the suspension, allow very cold liquids to be protected against external heat and to be stored for long times.

Chapter 8. Success Breeds Success

1. Part of the return on the public investment in space science is the release of news about significant discoveries. However, science and news operate on inconsistent time scales. In science, the emphasis is on repeated tests to ensure that a new finding is correct, but to be news something has to be announced immediately when it is discovered. Although some scientific discoveries are of the "eureka" kind whose significance is immediately apparent, the Mars life forms were far more subtle and controversial. Debate about whether the forms represent primitive life has continued long after the initial announcement.

To avoid presenting scientific results prematurely as firmly established conclusions, NASA frequently has a neutral or skeptical scientist take part in major releases. In the case of the Mars life, William Schopf, an authority on early life on Earth, took this role. He described the events as follows:

> I received a phone call from NASA Headquarters informing me that . . . Johnson Space Center scientists . . . had completed studies of a meteorite they claimed held evidence of ancient life on Mars. . . . [B]ecause some at headquarters thought the evidence "a bit iffy," they wanted an outside expert to publicly evaluate the findings when they were announced to the world. Would I, please, perform this task? I . . . [quoted] Carl Sagan's catchphrase that "extraordinary claims require extraordinary evidence."
>
> A few days later NASA called back and informed me that the agency's director, Dan Goldin, had personally pegged me for the job—partly, I gather, because he's a Sagan fan (and was said to have been pleased by the quote), but I think mostly because he knows it's in NASA's best interest to get the story straight. Any claim for life on Mars—whether of organisms small or large, past or present—is bound to stir controversy. This one would be no exception. The "iffy" evidence was certain to raise eyebrows, and since NASA's budget hearings were looming, even the timing of the announcement might be regarded as suspicious. . . . To protect NASA's reputation and at the same time stifle the easily predicted army of naysayers, he decided to assign a hard-nosed outsider to evaluate the claim.
>
> . . . [I]t seems to me that the Mars Meteorite Team . . . tackled a difficult interdisciplinary problem. An instant answer, pro or con, was not in the cards. The team's presentations were measured, sensible, their arguments plausible. By the time they finished, virtually the entire throng of journalists on hand seemed willing to believe. Introduced as the designated "skeptic" to "begin the debate," I had no doubt my words would prove unwelcome. I was like Daniel in the lion's den. But the evidence was (and still is) inconclusive, and it fell on me to point that out. Some claim the glass is half full; to others it's half empty. But no one who knows the facts would claim the glass is overflowing—not then, not now. Not even the Mars Meteorite Research Team. (Schopf 1999)

Chapter 15. 2000: The Rules Change

1. Although not officially a faster, better, cheaper mission, WIRE was built in a similar style.

Chapter 17. 2000: Selecting Some Science

1. A reprieve may be possible by continuing to use the two shortest-wavelength IRAC bands, whose detectors can operate well above liquid helium temperatures.

Chapter 26. Our First Year in Orbit

1. For a more up-to-date summary, visit the Spitzer Science Center Web site, http://www.spitzer.caltech.edu/Media/

BIBLIOGRAPHY

Augustine, N., et al. 1990. *Report of the Advisory Committee on the Future of the U.S. Space Program*. Washington, D.C.: National Aeronautics and Space Administration.

Bahcall, J. N. 1999. "Prioritizing Science: A Story of the Decade Survey for the 1990s." In *The American Astronomical Society's First Century*, ed. David DeVorkin. Washington, D.C.: Published for the American Astronomical Society through the American Institute of Physics.

Bahcall, J. N., et al. 1991. *The Decade of Discovery in Astronomy and Astrophysics*. Washington, D.C.: National Academy Press.

Bearden, D. 2001. "When Is a Satellite Mission Too Fast and Too Cheap?" Military and Aerospace Programmable Logic Devices (MAPLD) Conference, Johns Hopkins University.

Bossak, Brian. 2003. "Early Nineteenth-Century Hurricanes." Ph.D. diss., Florida State University, Tallahassee.

Canizares, A. 1999. "Recent Mars Foul-ups Cost Little Compared to Other NASA Missions." *Space.com*, 6 December.

Decin, G., et al. 2000. "The Vega Phenomenon around G Dwarfs." *Astronomy and Astrophysics* 357:533.

Field, G., et al. 1982. *Astronomy and Astrophysics for the 1980's*. Vol. 1. Washington, D.C.: National Academy Press.

Fisk, Lennard A. 2000. Interview. NASA Oral History Project. Obtained from NASA History Office.

Foley, T. 1995. "Mr. Goldin Goes to Washington." *Air and Space*, April–May.

General Accounting Office. 1994. *Space Projects: Astrophysics Facility Program Contains Cost and Technical Risks*. Report NSIAD-94-80. Washington, D.C.: U.S. Government Printing Office.

Habing, H., et al. 2001. "Incidence and Survival of Remnant Disks around Main-Sequence Stars." *Astronomy and Astrophysics* 365:545.

Huntress, W. T. 2003. Interview. NASA Oral History Project. http://www.jsc.nasa.gov/history/

"Mars Program Independent Assessment Team Summary Report." March 14, 2000. http://www.jpl.nasa.gov/marsreports/mpiat_summary.pdf

McCurdy, H. E. 2001. *Faster, Better, Cheaper: Low-Cost Innovation in the U.S. Space Program*. Baltimore: Johns Hopkins University Press.

Muirhead, Brian, and William Simon. 1999. *High Velocity Leadership: The Mars Pathfinder Approach to Faster, Better, Cheaper*. New York: HarperCollins.

Pritchett, Price, and Brian Muirhead. 1998. *The Mars Pathfinder Approach to Faster-Better-Cheaper: Hard Proof from the NASA/JPL Pathfinder Team on How Limitations Can Guide You to Breakthroughs*. Dallas: Pritchett and Associates.

Ray, Justin. 2003. Mission Status Center. *Spaceflight Now*, August 25.

Roman, Nancy G. 2000. Interview. NASA Oral History Project. http://www11.jsc.nasa.gov/history/

Roy, S. A. 1998. "The Origin of the Smaller, Faster, Cheaper Approach in NASA's Solar System Exploration Program." *Space Policy* 14:153.

Sarsfield, Liam. 1998. *The Cosmos on a Shoestring*. Santa Monica, Calif.: RAND.

Schopf, J. William. 1999. *Cradle of Life: The Discovery of Earth's Earliest Fossils*. Princeton: Princeton University Press.

Shirley, Donna, and Danelle Morton. 1998. *Managing Martians*. New York: Broadway Books.

Smith, Robert W., Paul A. Hanle, Robert H. Kargon, and Joseph N. Tatarewicz. 1993. *The Space Telescope: A Study of NASA, Science, Technology, and Politics*. Rev. ed. 1989. Cambridge: Cambridge University Press.

Space and Earth Science Committee, NASA Advisory Council. 1986. "The Crisis in Earth and Space Science." Washington, D.C.: National Aeronautics and Space Administration, p. iii.

Spaceflight Gossip. 2003. "Russia to Launch SIRTF." http://www.spaceflightgossip.com. 30 April.

"Space Science: Will NASA's Research Reforms Fly?" *Science*, November 17, 1995.

Spear, T. 2000. "NASA FBC Task Final Report." NASA, March 13.

Spitzer, L. 1989. "Dreams, Stars, and Electrons." *Annual Reviews of Astronomy and Astrophysics* 27:1.

Spitzer Science Center. 2004, 2005. Web site: http://www.spitzer.caltech.edu

Tucker, W., and K. Tucker. 2001. *Revealing the Universe*. Cambridge: Harvard University Press.

"WIRE Mishap Investigation Board Report." June 8, 1999. http://sunland.gsfc.nasa.gov/smex/wire/mission/updates/jun2599.html

ADDITIONAL READING

Books that Give Inside Views of Space Exploration

Beyond the Moon: A Golden Age of Planetary Exploration, 1971–1978, by Robert S. Kraemer (Washington, D.C.: Smithsonian Institution Press, 2000), is an overall look at the strategies behind the early NASA missions to the planets. More detail is provided for the Viking missions to Mars and the Voyagers to the outer planets. Kraemer writes in an accessible style about the engineering, financial, and political hurdles facing each mission, but includes relatively little about the science. He headed the planetary exploration program from 1970 until 1976.

The Challenger *Launch Decision: Risky Technology, Culture, and Deviance at NASA*, by Diane Vaughan (Chicago: University of Chicago Press, 1996), is by far the most thorough probe of the causes of the *Challenger* disaster. The material is organized to give an appreciation of how the fateful decision to launch looked from the before-launch perspective. The book shows how the disaster potential was underestimated and also provides substantial insight to how NASA works in general. The author of this highly recommended book is a sociology professor; the style is academic.

New Cosmic Horizons: Space Astronomy from the V2 to the Hubble Space Telescope, by David Leverington (Cambridge: Cambridge University Press, 2000), is a complete history of space science up to about 1997, particularly attractive for including American, European, Russian, and Japanese programs appropriately. The author, who managed a number of spacecraft projects in Europe, includes the technical and management challenges, plus discussions of major science results (although these are a bit uneven). For those with a solid understanding of basic astronomy, this book is highly recommended.

Revealing the Universe: The Making of the Chandra X-Ray Observatory, by Wallace Tucker and Karen Tucker (Cambridge: Harvard University Press, 2001). X-ray astronomer Wallace Tucker and his astronomer and writer wife, Karen, tell the story of getting the Chandra mission started and building it. Chandra is the third of the four Great Observatories, and many of the events in this book occurred in parallel with the story of Spitzer. The book is written in an accessible style and provides an interesting counterpoint regarding these events.

Rockets, Missiles, and Space Travel, 3d and rev. eds., by Willy Ley (New York: Viking, 1957, 1961), is a classic tale of early rocket development in Germany. The story starts with dedicated amateur visionaries (including Ley) with dreams of space travel.

Many of them helped build the V2, the first large high-performance rocket. The V2s and their developers then played central roles in postwar rocket development in the United States and USSR. Highly recommended and accessible to general readers.

The Space Telescope: A Study of NASA, Science, Technology, and Politics, rev. ed., by Robert W. Smith, Paul A. Hanle, Robert H. Kargon, and Joseph N. Tatarewicz (Cambridge: Cambridge University Press, 1989, 1993), is a history of the space telescope placed in the context of its role as a very big science project and the resulting strains on the management, political, and financial resources of NASA. The book discusses extensively the steps taken to "sell" the project and describes the technical challenges and the complex management issues among the NASA centers and aerospace contractors involved. Smith is a historian of science. This highly recommended book is well written in an academic style.

The Very First Light: The True Inside Story of the Scientific Journey Back to the Dawn of the Universe, by John C. Mather with John Boslough (New York: Basic Books, 1998), is a combination autobiography, explanation of theories about the early Universe, and history of the Cosmic Background Explorer (COBE). The book emphasizes how the team was organized to build this satellite, how they designed and built it, and how they interpreted the data from the mission. John Mather, the COBE project scientist, was assisted by writer John Boslough. The book is written in an accessible style and is highly recommended.

Excellent Books on Astronomy and Planet Exploration in General

Coming of Age in the Milky Way, by Timothy Ferris (Garden City, N.Y.: Anchor, 1989), is a classic on general astronomy; Ferris has written a number of other excellent books.

The Edge of Infinity, by Fulvio Melia (Cambridge: Cambridge University Press, 2003), describes massive black holes and their influence on the Universe; for general readers.

The Elegant Universe: Exploding Stars, Dark Energy, and the Accelerating Cosmos, by Robert Kirshner (Princeton: Princeton University Press, 2004), describes the discovery of dark energy and the accelerating expansion of the Universe.

The Elegant Universe: Superstrings, Hidden Dimensions, and the Quest for the Ultimate Theory, by Brian Greene (New York: Vintage, 2000), explains modern physics and its implications for theories of the Universe.

Jupiter Odyssey: The Story of NASA's Galileo Mission, by David M. Harland (New York: Springer-Verlag, 2000), is one of a number by the same author on many aspects of the manned and planetary exploration space programs.

Life Everywhere, by David Darling (New York: Basic Books, 2001), is an excellent short introduction to astrobiology.

A *Traveler's Guide to Mars*, by William K. Hartmann (New York: Workman Publishing, 2003). A planetary scientist and space artist takes you on a tour of Mars.

INDEX

ABOUT THE AUTHOR

George H. Rieke is Regents' Professor of astronomy and planetary science at the University of Arizona, where he began his career as a postdoctoral fellow in 1970. His involvement with Spitzer began in the 1970s and ran in parallel for many years with a career in groundbased infrared astronomy. He has been author or coauthor of more than 350 technical articles, as well as his first book, *The Detection of Light*. With his astronomer wife Marcia, he enjoys photography and travel.

ROWAN UNIVERSITY CAMPBELL LIBRARY
3 3001 00912 981 7